1st International Conference on Physical Problems of Engineering (1st ICPPE)

Journal of Physics: Conference Series Volume 2118

Bogota, Colombia
2 – 3 September 2021

ISBN: 978-1-7138-5089-2
ISSN: 1742-6588

Printed from e-media with permission by:

Curran Associates, Inc.
57 Morehouse Lane
Red Hook, NY 12571

Some format issues inherent in the e-media version may also appear in this print version.

Institute of Physics
Dirac House, Temple Back
Bristol BS1 6BE UK

Phone: 44 1 17 929 7481
Fax: 44 1 17 920 0979

techtracking@iop.org

Additional copies of this publication are available from:

Curran Associates, Inc.
57 Morehouse Lane
Red Hook, NY 12571 USA
Phone: 845-758-0400
Fax: 845-758-2633
Email: curran@proceedings.com
Web: www.proceedings.com

TABLE OF CONTENTS

1st International Conference on Physical Problems of Engineering (1st ICPPE)

J C Ruge[1], J E Torres-Ramírez[2], J Barba-Ortega[2,3], L M Dueñas[2], and E D V-Niño[2]
[1] Universidad Militar Nueva Granada, Bogotá, Colombia
[2] Foundation of Researchers in Science and Technology of Materials, Bucaramanga, Colombia
[3] Universidad Nacional de Colombia, Bogotá, Colombia

E-mail: info@foristom.org

The reciprocal contribution that physics has had in engineering and engineering in physics is undeniable. The First International Conference on Physical Problems of Engineering (1st ICPPE) tries to highlight this symbiotic relationship between both sciences through different presentations that reflected their results in separate articles that will be published in the Journal of Physics: Conference Series.

Five keynote lectures invited from Spain, Portugal, the United States of America, Brazil, and Colombia, took place on different topics allowed the audience to learn about the current state of engineering knowledge within the physics proposed in the event. The pandemic did not prevent the remote virtual assistance of students, professors, researchers, and the entire interested academic and scientific community; the questions and discussions in a respectful environment further enriched the contents of the event and the socialization of the research results itself. The website of the conference is available at https://foristom.org/1icppe.

The organizing committee 1st ICPPE thanks all authors who presented their contributions in this edition of the event—the success of this meeting more than guarantees the second meeting in 2022, hopefully in face-to-face conditions, where we can discuss concepts in a different environment. Besides, we are proud to have carried out this event successfully and increased the generation of new knowledge regarding the event's central theme. We will be waiting for you at the 2nd ICPPE in 2022.

On behalf of the organizing committee of the 1st ICPPE, we are extremely thankful to all authors and participants for providing their valuable contributions to this Proceedings volume as well as the reviewers for their constructive recommendations and criticism aiding to improve the presented articles. Likewise, we like to thank the Universidad Militar Nueva Granada, and the Foundation of Researchers in Materials Science and Technology, for all the support technical and logistical received.

Finally, the editor hopes that those interested in the area of physical problems of engineering can enjoy this reading, of the volume of the Journal of Physics: Conference Series (JPCS), which reflects a wide variety of current issues.

"Education and expression freedom for anyone".

Journal of Physics: Conference Series 2118 (2021) 011001 doi:10.1088/1742-6596/2118/1/011001

Organizing Committee

Juan Carlos Ruge (Chairman)

Oscar Javier Reyes
Julián Carrillo León
Javier Camacho Tauta

National Scientific Committee

- Dr. Gilberto Martínez Arguelles
  Universidad del Norte
- Dr. Martin Estrada
  Universidad Nacional de Colombia
- Dr. Julio E. Colmenares
  Universidad Nacional de Colombia
- Dr. José José Barba
  Universidad Nacional de Colombia
- Dr. Jorge A. Pineda
  Universidad Distrital Francisco José de Caldas
- Dr. Juan Gabriel Bastidas
  Universidad Católica de Colombia
- Dr. Wilfredo Marimón
  Universidad Católica de Colombia
- Dr. Alfonso M. Ramos
  Pontificia Universidad Javeriana
- Dr. Ivan F. Otalvaro
  Pontificia Universidad Javeriana
- Dr. Eliana Martínez Rojas
  Universidad Piloto de Colombia
- Dr. Carlos Felipe Urazán
  Universidad de La Salle
- Dr. Adela Tatiana Rodríguez
  Universidad Militar Nueva Granada
- Dr. Hebert G. Rivera
  Universidad Militar Nueva Granada
- Dr. Ely Dannier V. Niño
  Foundation of Researchers in Science and Technology of Materials

Journal of Physics: Conference Series **2118** (2021) 011001 doi:10.1088/1742-6596/2118/1/011001

International Scientific Committee

- Dr. Carlos José Slebi
  Universidad de Cantabria, Spain
- Dr. Renato P. da Cunha
  Universidad de Brasilia, Brazil
- Dr. Paulo J. R. Albuquerque
  Universidad de Campinas, Brazil
- Dr. Jorge Cormane
  Universidad de Brasilia, Brazil
- Dr. (c) Fausto Molina-Gomez
  Universidad de Porto, Portugal
- Dr. (c) Anderson Sandoval Amador
  Universidad Nacional del Centro de la Provincia de Buenos Aires, Argentina

Sponsor

- ✓ Universidad Militar Nueva Granada

Partners

- ✓ Grupo de Investigación Estructuras y Sísmica, Universidad Militar Nueva Granada
- ✓ Grupo de Investigación Geotecnia, Universidad Militar Nueva Granada
- ✓ Foundation of Researcher in Science and Technology of Materials
- ✓ IOP Publishing

Journal of Physics: Conference Series **2118** (2021) 011001 doi:10.1088/1742-6596/2118/1/011001

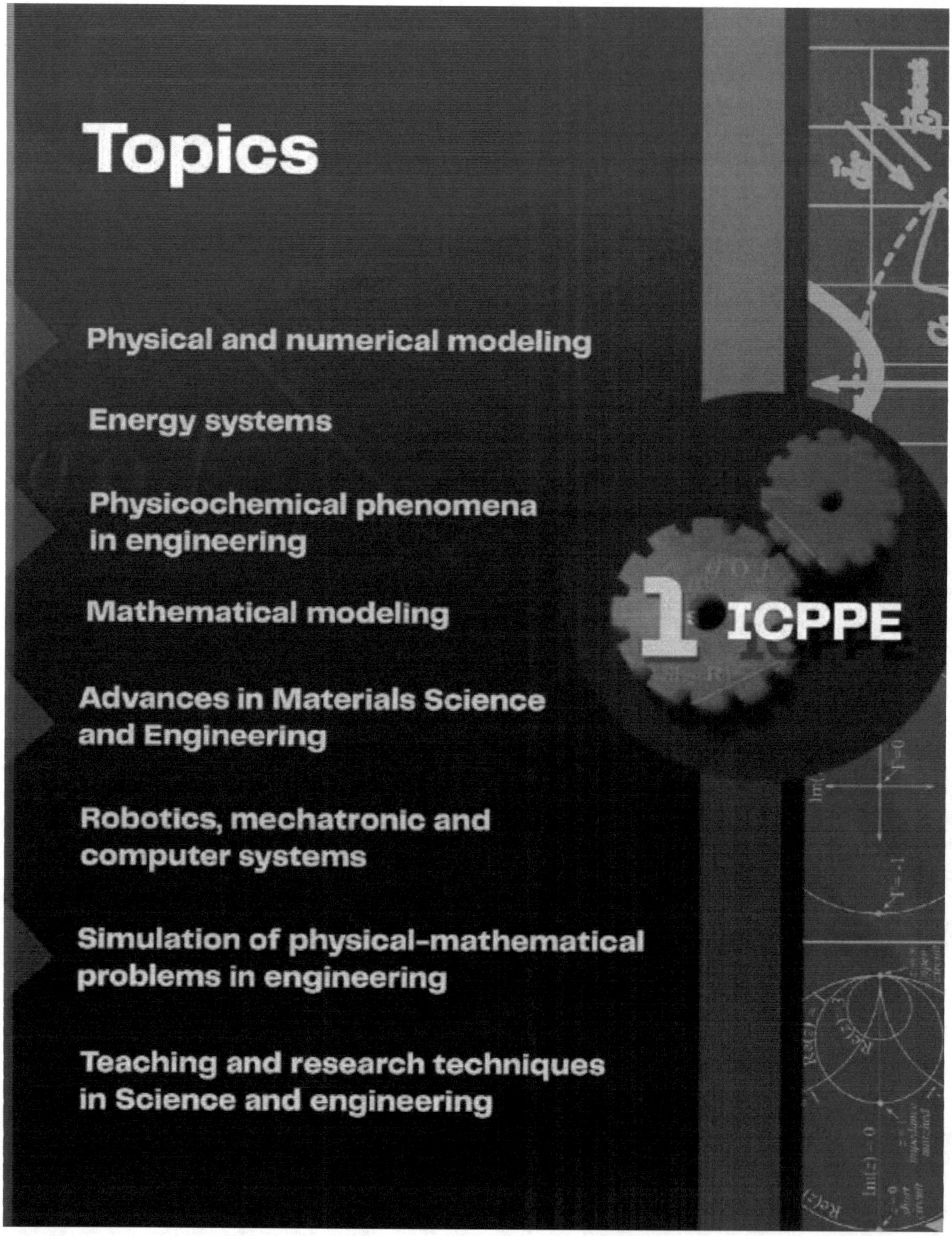

Journal of Physics: Conference Series **2118** (2021) 011001 doi:10.1088/1742-6596/2118/1/011001

Journal of Physics: Conference Series **2118** (2021) 011002 doi:10.1088/1742-6596/2118/1/011002

Peer review declaration

All papers published in this volume of Journal of Physics: Conference Series have been peer reviewed through processes administered by the Editors. Reviews were conducted by expert referees to the professional and scientific standards expected of a proceedings journal published by IOP Publishing.

- **Type of peer review:** Single blind

- **Conference submission management system:**
  By email received from the organizing and editorial committee of the conference. The correspondence authors make the submission by email.

- **Number of submissions received:**
  23

- **Number of submissions sent for review:**
  23

- **Number of submissions accepted:**
  21

- **Acceptance Rate (Number of Submissions Accepted / Number of Submissions Received X 100):**
  91.30%

- **Average number of reviews per paper:**
  2 reviews per paper

- **Total number of reviewers involved:**
  10

- **Any additional info on review process:**
  All papers are plagiarism checked by Turnitin software

- **Contact person for queries:**
  Ely Dannier V. Niño
  Foundation of Researchers in Science and Technology of Materials (FORISTOM)
  Email: info@foristom.org

Journal of Physics: Conference Series **2118** (2021) 012001 doi:10.1088/1742-6596/2118/1/012001

Temporal analysis of the heavy metal concentration in road sediment and dust using statistical models

C A Zafra-Mejía[1], H A Rondón-Quintana[1], and L C Echeverry-Prieto[1]
[1] Facultad del Medio Ambiente y Recursos Naturales, Universidad Distrital Francisco José de Caldas, Bogotá, Colombia

E-mail: czafra@udistrital.edu.co, harondonq@udistrital.edu.co

Abstract. The objective of this paper is to show a temporal analysis using autoregressive integrated moving average models of the heavy metal concentration in road sediment and dust of Soacha, a Colombian locality. The representative size fractions in the road sediment and dust were <250 μm and ≤10 μm, respectively. The results suggest that lead is the best metallic element to study the relationship between the heavy metal concentration in the road sediment and dust (r-Pearson = 0.90). Univariate models ($R^2 \geq 0.58$) suggest that the time series of lead concentrations in road sediment and dust have the same temporal structure. Namely, because they are first-order autoregressive processes, concentrations at a given moment of time are influenced by the immediately preceding concentrations. The transfer function model ($R^2 = 0.91$) suggests that there is no delay in impulse transfer from road dust concentration to lead concentration in the road sediment. The effect is immediate for a sampling interval of 3 days. The results show that modeling has a better fit during the rainy season compared to the dry season. In the context of the simulation of physical phenomena in engineering, this study is relevant to deepen knowledge in relation to the use of autoregressive integrated moving average models.

1. Introduction

Urban surface pollution with heavy metals (HMs) required special attention due to its toxic effects on human health [1]. Road sediments (RSs) accumulated in dry time associated HMs that affected water resources when they were transported by surface runoff [2]. In addition, RSs affected the air quality when were re-suspended by wind and traffic-induced turbulence. This re-suspended material was commonly identified as road dust (RD) [3].

The available literature on methodologies for studying the HMs associated with RS from information on air quality or RD is scarce. The RS has been recognized as an important urban source of atmospheric particulate matter ≤10 μm (PM_{10}) [4]. This contribution of particulate atmospheric material was due to the mechanical disintegration of the RS by crushing, grinding, and explosion processes [5]. Thus, exposure to atmospheric particulate material generated on roads involved increased exposure to ultrafine particles and PM_{10} emissions from vehicle exhaust, and from brake and tire wear [6]. Therefore, it is then necessary to study the HM content in RS and RD from new approaches to analysis. In this way, the time-series analysis using autoregressive integrated moving average (ARIMA) models can be useful to deepen knowledge in relation to the phenomenon of HM accumulation in RS and RD.

The ARIMA models [7] were effectively used in the study of HM content in different environmental matrices [8] and were widely used for their simplicity and statistical characteristics [9]. These models also offer an effective method for obtaining accurate outcomes without considerable calibration time.

ARIMA models examine the probable temporal association between present and previous information and its errors and shoulder a linear relationship between the data. Hence, each observation can be described as a linear function of its previous data with an extra error term [10].

Explicitly, these models consider the information sequence over time as a random series and analyze a fraction of the information in the sequence to calculate definite parameters that specify the mathematical model, and then model the time series. The ARIMA models consider the residual information in the sequence to validate the mathematical model, and the validated model is utilized to forecast successive data in the time series [11].

The objective of this paper is to show a temporal analysis with ARIMA models of the HM concentration (Pb, Zn, Cu, Ba, and Fe) associated with RS and RD in Soacha,a Colombian locality. This research will be relevant to deepen knowledge in relation to the following aspects in the context of the simulation of physical phenomena in engineering: (i) the temporal behavior of HMs associated with RS, (ii) the possible relationship between HM concentrations in RS and RD, and (iii) the use of ARIMA models to study the temporal relationship of HMs in RS and RD.

2. Materials and methods

This chapter shows a description of the study site and sampling system, and the protocols for laboratory and data analysis.

2.1. *Study site*

The study road was in the municipality of Soacha, Colombia, in central Colombia, 4°35'05" N, 74°13'12" W. Its tropical mountain climate (cold climate) was characterized by an average annual temperature of 14 °C, wide daily temperature variation (up to 18.0 °C), average annual relative humidity of 66.0%, and average annual rainfall of 900 mm; all climatological data were obtained from a station located at a 340 m distance from the road surface under study. The study road was in the urban center of Soacha, Colombia, adjacent to the central municipal hospital. The road basin had an average elevation of 2560 m.a.s.l., average slope of 1.70%, and waterproof coverage in 90% of its area. The road surface had a one-way street with two lanes for vehicle traffic. The road gives access to residential and institutional areas but was occasionally used by commercial vehicles. In order of importance the land use in the study area was residential, institutional, and commercial.

2.2. *Sampling system*

The RS collection system consisted of three components: (i) a wooden frame with an area of 0.50 m^2 (0.707 m x 0.707 m), (ii) a manual brush of synthetic fibers, and (iii) a manual plastic collector [12]. Samples of RS were taken in dry time (every 3 days), at 0.40 m from the curb, between 11:00 - 13:00 hours, and over a continuous period of 127 days (07/01/2018-14/05/2018). The road surface was swept with a brush of synthetic fibers (3 repetitions) so that the sediments most strongly adhered to it were available to be collected. The total number of samples collected on the road surface was 44.

The collected RS was stored in airtight plastic bags to later be taken for lab analysis. The sampling protocol for total suspended particle (TSP) and PM_{10} (RD) was based on the U.S. Environmental Protection Agency (EPA/625/R-96/010a-IO-2.1) [13]. The TSP sampling system consisted of high-volume manual equipment with mass flow control (Thermo Environmental Instruments Inc., GMW TSP high volume air sampler - mass flow). The sampling system for PM_{10} consisted of a high-volume manual cyclone-type equipment with volumetric flow control (Wedding & Associates, 600 PM_{10} critical flow high volume air sampler - volumetric flow).

2.3. *Lab analysis*

The particle size distribution of RS (< 63 μm - 2800 μm) was determined from ISO 2591-1 standard [14]. The HM extraction from RS and RD was carried out from the ISO 11466 method [15]. The method used to determine the HM content (Pb, Zn, Cu, Ba, and Fe) in RS and RD was as follows: SM 3125 B, inductively coupled plasma-mass spectrometry (ICP-MS) Method [16]. The selection, preparation, and gravimetric determination method for TSP and PM_{10} filters was based on the U.S. Environmental Protection Agency (EPA/625/R-96/010a-IO-3.1) [13].

2.4. Data analysis
The data analysis considered two stages; stage 1: A descriptive statistical assessment was applied to identify anomalies in HM concentration time series. The normal distribution of HM concentration time series was estimated by a Kolmogorov/Smirnov test (p-value > 0.05). Associations between HMs were examined using Pearson's correlation coefficient (r) to identify probable control HMs of their behavior in RS and RD. In this study, we supposed that the HMs with the best associations were those that perhaps described their behavior in RS and RD. Lastly, the influence of rainfall on the HM concentrations in RS and RD was studied. Thus, two rainfall periods were studied: dry season or decrease in rainfall (January/07-February/08) and rainy season (February/09-May/14).

Stage 2: ARIMA models for HM concentration time series were generated applying the methodology of Box, *et al.* [7]. The three phases stated by Box, *et al.* [7] were considered in the construction of the models: identification/estimation/verification. During the model identification phase, the orders of the autoregressive and moving average polynomials were calculated, as well as the degree of differentiation to eliminate the non-seasonality of the time series under study. Specifically, the orders 'p', 'd', and 'q' of the models were calculated. In the second phase, the parameters of the autoregressive and moving average polynomials were calculated by the maximum likelihood method for each model found.

The eight hypotheses suggested by Box, *et al.* [7] were checked to choose the best models for each time series. The appropriateness of modelling was determined using the Ljung and Box (Q') parameter. A p-value > 0.05 meant that the model was appropriately generated to explain the association information in time series [17]. Another parameter applied to distinguish the best model was the normalized Bayesian information criterion (normalized BIC). The best model was chosen when the BIC was the lower [18]; All data analysis was implemented using the IBM-SPSS V.18.0 software;

3. Results and discussion
The following are the results and discussions related to the HM concentration in RS and RD, and to the ARIMA models developed.

3.1. Heavy metal concentration in road sediment and road dust
The finer fraction of the collected RS (< 63 μm) showed the highest HM concentrations (Table 1). Overall, the HM concentration tended to decrease as the particle size of the RS increased. When comparing the HM concentrations between PM_{10} and TSP, it was observed that these were higher in the potentially respirable fraction (PM_{10}). For all study HMs, the concentrations associated with PM_{10} were 3.09 times higher than TSP (Table 2). Therefore, the results suggested that the HM sources mainly emitted particles of size less than or equal to 10 μm.

Chen et al. [4] reported similar findings. Pearson's correlation analysis was performed between HM concentrations in the RS (< 250 μm) and PM_{10}. The results showed that Pb was the HM that showed the best correlation coefficient (r = 0.90), followed by Cu (0.66), Ba (0.62), Fe (0.40), and Zn (0.12). The HMs with the best correlation coefficients were possibly best suited to study the relationship between HM concentrations in the RS and PM_{10} (Pb > Cu > Ba). In this way, Pb was selected as the representative element to study the relationship between the HM concentration in RS and PM_{10} (RD).

3.2. Autoregressive integrated moving average analysis
Univariate ARIMA models were developed for the time series of Pb concentrations in the RS and PM_{10} (RD). The time series of PM_{10} concentration showed a decreasing trend during the study period. This decrease was possibly associated with rainfall behavior. Namely, in the dry season (January/07 - February/08) the highest PM_{10} concentrations were observed while in the rainy season (February/09 - May/14) the concentrations tended to decrease. In addition, it was observed that the variation in PM_{10} concentration showed a pattern of decrease. In the dry season, the variation in the PM_{10} concentration

was greater than in the rainy season. The time series for the Pb concentration in RS showed the same trends (Figure 1). Thus, the results suggested that the time series under study did not correspond to stationary processes.

Table 1. HM concentration according to the size fraction of RS (95% confidence interval).

HM	Size fraction (μm)						
	< 63	63-125	125-250	250-500	500-1000	1000-2000	2000-2800
Pb [a]	99±26	79±21	74±19	81±21	56±15	44±11	32±80
Zn	137±19	86±12	65±90	57±80	29±40	18±30	7±20
Cu	46±11	39±90	38±90	33±80	17±40	12±30	10±20
Ba	192±46	139±33	105±25	90±22	47±11	26±60	12±30
Fe [b]	11.69±0.9	8.79±0.7	8.25±0.7	9.60±0.8	7.45±0.6	4.59±0.4	2.74±0.2

*HM concentration in mg/kg of dry matter. [a] HM; [b] concentration in g/kg of dry matter.

Table 2. HM concentration in RD: PM_{10} and TSP (95% confidence interval).

HM	Size fraction	
	PM_{10} ($\leq$ 10 μm)	TSP ($\leq$ 100 μm)
Pb	1818±473	864±184
Zn	16312±5256	3065±1201
Cu	1197±273	458±700
Fe	10515±2688	9331±1766
Ba	7001±2247	1625±5730

*Concentration in μg/g of dry matter.

From the viewpoint of the ARIMA process, it was observed that the time series of Pb concentration in the RS had a short memory. Namely, being a first-order autoregressive process (AR-p = 1), it was observed that the Pb concentration in the RS was influenced by the concentration immediately preceding (Table 3). In addition, the time series of PM_{10} concentration showed a similar ARIMA structure (AR-p = 1). Thus, it was probably a phenomenon of short memory as well as the phenomenon of Pb concentration in the RS. Vicente, *et al.* [19] obtained similar results by studying daily air pollution by particulate matter, NO_2, As, Cd, Ni, and Pb. These researchers identified a short-memory ARIMA model, that is, with a second-order autoregressive term (AR-p = 2).

As observed, there was the possibility of representing the two-time series under study (Pb and PM_{10}) using the same ARIMA time structure (1,0,0). It is important to mention, that in this study the immediately preceding observation corresponded to three days. This is based on the frequency of RS collection and PM_{10} sampling. In the ARIMA models developed, a moving average term (MA-q = 0) was not evidenced. In other words, no fluctuations were visualized in the time series under study. This with respect to the average value of Pb and PM_{10} concentrations observed during the study period.

Table 3. Univariate ARIMA models for the time series of Pb concentrations in RS and PM_{10} (RD).

Time series	Model (p,d,q)	Transf.	R^2	RSME	MAPE	MAE	Q´ p-value	BIC	Outliers
Pb	(1,0,0)	LN	0.577	34.17	32.01	23.69	0.577	7.313	1
PM_{10}	(1,0,0)	NT	0.831	10.45	30.78	8.05	0.944	5.028	2

*Transf. = transformation; LN = natural logarithm; NT = no transformation. When a transformation was applied, the generated model included a constant.

After analyzing the univariate models, a transfer function model was developed to relate the two-time series (Pb and PM_{10}) using an ARIMA model. The hypothesis for the development of the model was the probable existence of a relationship between the Pb concentrations in RS and PM_{10} (RD). Namely, it could be expected that there was a dynamic relationship between the two variables under study. Thus, a unidirectional causal relationship was suggested from the PM_{10} concentration to the Pb concentration in RS. The transfer function model considered a single input and output, corresponding to a system related by a linear filter as proposed by Box, *et al.* [7].

The transfer function modeling allowed the identification and estimation of the transfer function and the noise model, based on the information provided by the time series of PM_{10} (independent variable) and Pb (dependent variable) concentrations in RS. The univariate models obtained for the input variable (PM_{10} concentration) were used to perform the pre-bleaching of the output variable (Pb concentration in RS). This to determine the transfer function [20].

The estimated cross-correlation function between the two study variables was also calculated to determine the impulse response function. The results showed that the '0' delay was the only one statistically significant, according to the limit line obtained using the Bartlett criterion [21]. Namely, in this study, there was no delay in the impulse transfer from the PM_{10} concentration to the Pb concentration in RS. It is important to mention that the PM_{10} monitoring station was located 5.0 m from the curb of the roadway where the RS was collected. Once the transfer function was identified, the noise series was estimated, and the ARMA process generating the residues was identified through its simple and partial autocorrelation functions.

The results showed that the residues followed an ARMA process (0,0). In addition, an outlier was identified that was included in the model of the complete transfer function (dummy variable). The outlier was in data 12 (February/03). This data corresponded to a minimum value in the concentrations of PM_{10} (RD) and Pb (RS), which preceded the maximum concentrations of the variables under study, respectively (Figure 1). This outlier was of the innovative type, that is, in the time series it affected each observation from a specific starting point in the time series.

The statistics of the transfer function model were studied to evaluate the goodness of fit in time series modeling (Table 4). In this regard, an estimate was made of the proportion of total variance in the series explained by the model ($R^2 = 0.910$). Thus, the PM_{10} concentration explained 91.0% of the variation in the Pb concentration in RS. The results showed that the adjustment of the transfer function model was better compared with the adjustment obtained for each of the univariate models of the time series of concentrations of PM_{10} ($R^2 = 0.831$) and Pb ($R^2 = 0.577$).

Additionally, the average absolute error (MAE) of the ARIMA modeling showed that the predicted values for the time series of Pb concentration in the RS deviated on average 11.36 mg/kg with respect to the observed values (Table 4). The average absolute percentage error (MAPE) of the modeling showed an average deviation of 15.2% between the observed and predicted values for the Pb concentration in RS. In order to detect the worst case in the prediction, we proceeded to determine the maximum mean absolute error (MaxAE: 52.1 mg/kg; MaxAPE: 53.6%).

The results suggested that the maximum errors in the forecast tended to occur in the dry season (*e.g.*, February/10), and the smallest errors tended to occur in the rainy season (*e.g.*, May/10) (Figure 1). This from the periods identified with respect to rainfall (dry season and rainy season). As is known, the highest Pb concentrations in RS occurred in the dry season, and the lowest concentrations occurred in the rainy season. Thus, the ARIMA model of transfer function developed tended to show a better fit during the rainy season.

Table 4. ARIMA model of transfer function between the concentrations of PM_{10} (RD) and Pb (RS).

Variable	Time series	Model (p,d,q)	Transf.	R^2	RSME	MAPE	MAE	Q´ p-value	BIC	Outliers
Dependent	Pb	(1,0,0)	LN	0.910	16.17	15.21	11.35	0.795	5.983	1 (Innovative)
Independent	PM_{10}	(0,0,0)	NT							

*Transf. = transformation; LN = natural logarithm; NT = no transformation. When a transformation was applied, the generated model included a constant.

Finally, the confidence limits (95%) for the predictions of Pb concentration in RS were determined to evaluate possible scenarios in the time series (Figure 1). The results showed that in the dry season the highest Pb concentrations were observed in RS. From the upper limit of the confidence interval, the concentration could reach values of up to 378 mg/kg. Namely, in dry weather there was a probability of reaching concentrations of up to 4.68 times above the average Pb concentration for the entire study period (80.7 mg/kg). Indeed, in the rainy season the lowest values in the Pb concentration in RS were observed. From the lower limit of the confidence interval, Pb concentration in RS could reach values of up to 20.2 mg/kg. Thus, in rainy weather there was a probability of reaching concentrations of up to 4.0 times below the average Pb concentration observed during the study period.

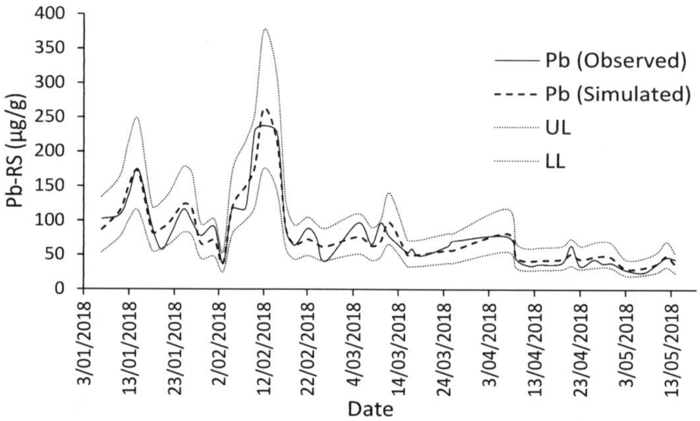

Figure 1. Pb concentrations observed and simulated with the ARIMA transfer function model. UL = upper limit and LL = lower limit (95% confidence interval).

4. Conclusions

The results of this study on the HM concentration in RS and RD using ARIMA models allow to visualize the following conclusions.

The results suggest that Pb is the best element to study the relationship between HM concentration in the RS and PM_{10} ($r = 0.90$). Overall, there is a coincidence in the correlations between HMs in the RS and PM_{10} (RD). In this regard, Pb, Cu, and Ba are suggested as possible indicators to study the behavior of HMs in road environments (RS and RD).

The univariate ARIMA models ($R^2 \geq 0.58$) suggest that the time series of Pb concentrations in the RS and PM_{10} have the same temporal structure. These two-time series are short memory. Namely, because they are first-order autoregressive processes (AR = 1), the concentrations at a given instant of time are influenced by the immediately previous concentrations. In this study, the analysis interval was three days. According to the ARIMA approach, no fluctuations are observed in the time series under study (MA = 0).

The ARIMA model of transfer function suggests that there is no delay in the impulse transfer from the PM_{10} concentration to the Pb concentration in RS. Namely, the effect is immediate from the analysis time interval considered in this study (3 days). The transfer function model shows a good fit ($R^2 = 0.910$). The average absolute percentage error of the modeling is 15.2%. The results suggest that ARIMA modeling has a better fit during the rainy season compared to the dry season. Indeed, the innovative atypical data detected occurs during the dry season.

From the confidence limits generated by ARIMA modeling, it is observed that during the dry season the Pb concentrations in RS can be up to 4.68 times higher compared to the average concentration value calculated for the whole study period. In the rainy season, Pb concentrations can be up to 4.0 times lower compared to the calculated average concentration value.

Finally, this study is relevant to deepen knowledge in relation to the use of ARIMA models for the simulation and interpretation of physical phenomena in engineering, in this specific case, associated with environmental and earth sciences.

References

[1] Kaonga C C, Kosamu I B M, Utembe W R 2021 A review of metal levels in urban dust, their methods of determination, and risk assessment *Atmosphere* **12** 891

[2] Jeong H, Choi J Y, Lee J, Lim J, Ra K 2020 Heavy metal pollution by road-deposited sediments and its contribution to total suspended solids in rainfall runoff from intensive industrial areas *Environmental Pollution* **265** 115028

[3] Aguilera A, Bautista F, Gutiérrez-Ruiz M, Ceniceros-Gómez A E, Cejudo R, Goguitchaichvili A 2021 Heavy metal pollution of street dust in the largest city of Mexico, sources and health risk assessment environ monit Assess **193** 193

[4] Chen S, Zhang X, Lin J, Huang J, Zhao D, Yuan T, Huang K, Luo Y, Jia Z, Zang Z, Qiu Y, Xie L 2019 Fugitive road dust PM2.5 emissions and their potential health impacts *Environ. Sci. Technol.* **53** 8455

[5] Amato F, Pandolfi M, Viana M, Querol X, Alastuey A, Moreno T 2009 Spatial and chemical patterns of PM10 in road dust deposited in urban environment *Atmospheric Environment* **43** 1650

[6] Schauer J J, Lough G C, Shafer M M, Christensen W F, Arndt M F, DeMinter J T, Park J-S 2006 Characterization of metals emitted from motor vehicles *Res. Rep. Health Eff. Inst.* **133** 77–88

[7] Box G E P, Jenkins G M, Reinsel G C, Ljung G M 2016 *Time Series Analysis: Forecasting and Control* (New Jersey: John Wiley & Sons)

[8] Romero-Barreiro M del P, Pinilla-Castañeda R D, Zafra-Mejía C A 2015 Evaluación temporal de la concentración de metales pesados (Pb y Cu) asociada con el sedimento vial: Fontibón-Barrios Unidos (Bogotá D. C., Colombia) *Ingeniería y Universidad* **19** 69

[9] Nourani V, Asghari P, Sharghi E 2021 Artificial intelligence based ensemble modeling of wastewater treatment plant using jittered data *Journal of Cleaner Production* **291** 125772

[10] Dellana S A, West D 2009 Predictive modeling for wastewater applications: Linear and nonlinear approaches *Environmental Modelling & Software* **24** 96

[11] Valipour M 2015 Long-term runoff study using SARIMA and ARIMA models in the United States *Meteorological Applications* **22** 592

[12] Mejía C Z, González J T, Monzón I T 2011 Concentración y distribución de metales pesados (Pb, Zn, Cu, Cd y Cr) en sedimentos viarios urbanos *Revista Facultad de Ingeniería Universidad de Antioquia* **58** 53

[13] United States Environmental Protection Agency (U.S. EPA) 2019 *Compendium of Methods for the Determination of Inorganic Compounds in Ambient Air, EPA/625/R-96/010a* (United States of America: United States Environmental Protection Agency)

[14] International Organization for Standardization (ISO) 2007 *Test Sieving - Part 1: Methods using Test Sieves of Woven Wire Cloth and Perforated Metal Plate, ISO 2591-1* (Switzerland: International Organization for Standardization)

[15] International Organization for Standardization (ISO) 2007 *Soil Quality - Extraction of Trace Elements Soluble in Aqua Regia, ISO 11466* (Switzerland: International Organization for Standardization)

[16] American Public Health Association (APHA) 2018 *Metals by Inductively Coupled Plasma-Mass Spectrometry, Standard Methods for the Examination of Water and Wastewater, SM 3125* (United States of America: American Public Health Association)

[17] Ljung G M, Box G E P 1978 On a measure of lack of fit in time series models *Biometrika* **65** 297

[18] Schwarz G 1978 Estimating the dimension of a model *The Annals of Statistics* **6** 461

[19] Vicente A B, Jordán Vidal M M, Sanfeliu Montolio T, Sánchez Á, Esteban M D 2012 Air pollution prediction models of particles, As, Cd, Ni and Pb in a highly industrialized area in Castellón (NE, Spain) *Environmental Earth Sciences* **66** 879

[20] Pankratz A 1991 *Forecasting with Dynamic Regression Models* (New Jersey: John Wiley & Sons)

[21] Bartlett M S 1946 On the theoretical specification and sampling properties of autocorrelated time-series *Journal of the Royal Statistical Society* **8** 27

Flow hydraulic simulation through two sand traps, using Ansys fluent

C D Rodríguez[1], **and J S De Plaza**[1]
[1] Facultad del Medio Ambiente y Recursos Naturales, Universidad Distrital Francisco José de Caldas, Bogotá, Colombia

E-mail: cdrodriguezb@correo.udistrital.edu.co, sjsebastiand@correo.udistrital.edu.co

Abstract. Computational fluid dynamics is a tool that allows to simulate and observe the behavior of any fluid, based on a physical, hydraulic, and hydrodynamic analysis. This research analyses the behavior of the flow in a sand trap, which is a structure used to remove sand particles with a minimum size of 0.10 mm, prior to treatment in a drinking-water plant. The objective of this study is to determine the highest efficiency between two sand traps, one with a double smooth screen and the other with a double perforated screen (with diffusers), based on the simulation and analysis behavior of the flow inside each sand trap. The methodology used includes the traditional design of each unit based on Hazen's model and Stokes viscosity law, to later carry out the numerical model simulation from Ansys Fluent (pre-processing, processing, and post-processing). The result shows that perforated double screen sand trap generates a removal efficiency of 78%, while the smooth double screen 28%. In addition, other four units of interleaved screens are proposed, in these cases efficiencies of up to 50% are observed and it is shown that it is necessary to implement at least two perforated screens (with diffusers) to guarantee an efficiency greater than 70%. Hydraulic simulation has a broad impact on infrastructure works and consulting.

1. Introduction

One of the fundamental structures implemented in drinking water pre-treatment, is the sand trap (ST), this structure allows to carry out the solids' removal of a certain diameter, as required, in order to facilitate the treatment in the plant [1]. This hydraulic type of structure can be simulated using computational fluid dynamics (CFD) [2], through this tool it is possible to obtain an estimated visualization of the flow behavior inside the ST and hydraulic operation analysis of the ST [3]. ANSYS is a software specialized in simulation, and within its components it is possible to perform a high precision CFD type simulation [4].

In this way, it is proposed to carry out the evaluation with two different types of ST, one with a double perforated screen and the other with a double smooth screen, and thus determine the maximum efficient in removing sand [5].

The determination of the best design depends on the analysis of the physical variables and the flow behavior inside the ST, however, the determining factor is the one that measures the removal of solids [6], since this is the one that defines the ST efficiency. Likewise, it is essential to consider all the physical phenomena present, such as velocity flow and turbulence to implement a structure with high removal efficiency [3,6]. Determining which unit, ST, is more efficient, prior to its construction, allows optimizing costs and foreseeing improvements for future problems.

2. Methodology and materials

The applied methodology consists in design development, then elaborate the 3D structures modeling and later, moving on to the development stages in the software, which are pre-processing, processing and post-processing, and finally get the data and analyze it.

2.1. Sizing and 3D model

The ST sizing is made from Stokes Law and the application of Allen Hazen`s model, as appropriate [7]. These models are based on the particle`s sedimentation rate and the unit (ST) efficiency [8,9]. See the Equation (1) and Equation (2).

$$\mathrm{Vs} = \frac{g(\rho_s - \rho_w)}{18*g} * d_s^2. \tag{1}$$

$$\%\mathrm{Ef.} = -0.0374\mathrm{Hz}^6 + 0.7669\mathrm{Hz}^5 - 6.2939\mathrm{Hz}^4 + 26.858\mathrm{Hz}^3 - 66.116\mathrm{Hz}^2 + 101.76\mathrm{Hz}, \tag{2}$$

where Vs is sedimentation velocity, g is gravity acceleration, ρ_s is particle density, ρ_w is fluid density and d_s is particle diameter. With the dimensions obtained and considering what is stipulated mainly for the analysis, the design is carried out in three dimensions using AutoCAD software, applying the tools indicated for 3D modeling. Once the model is elaborate in AutoCAD, it is exported to a sat format, since in ANSYS this format turns out to be compatible to perform the analysis.

2.2. Pre-processing

In this section, the analysis volume is defined from the generated geometry, from which the model in study is meshed. Then the relevant mathematical model is produced, which in turn is composed of the turbulence, discrete phase, and flow motion models.

2.2.1. Defining the analysis volume from the geometry and meshing. The analysis volume is defined from the geometry, flow limits, and the faces that function as contact surfaces. Subsequently, the meshing is done defining a cell size of 10 mm and tetrahedral type for greater precision, this also confers a more stable model from the courant number [9].

2.2.2. Turbulence model. The K-epsilon model is implemented like a turbulence model; this model represents the flow turbulence and turbulence dissipation properties through Equation (3) and Equation (4) [10], which are defined as follows.

Turbulent kinetic energy:

$$\frac{\partial}{\partial t}(\rho k) + \frac{\partial}{\partial_{xi}}(\rho k u_i) = \frac{\partial}{\partial_{xj}}\left[\left(\mu + \frac{\mu t}{\sigma_\varepsilon}\right)\frac{\partial k}{\partial x_j}\right] + G_k + G_b - \rho_\varepsilon - Y_m + S_k. \tag{3}$$

Turbulent dissipation:

$$\frac{\partial}{\partial t}(\rho\varepsilon) + \frac{\partial}{\partial_{xi}}(\rho\varepsilon u_i) = \frac{\partial}{\partial_{xj}}\left[\left(\mu + \frac{\mu t}{\sigma_\varepsilon}\right)\frac{\partial \varepsilon}{\partial x_j}\right] + C_{1\varepsilon}\frac{\varepsilon}{k} + (G_k + C_{3\varepsilon}G_b) - C_{2\varepsilon}\rho\frac{\varepsilon^2}{k} + S_\varepsilon, \tag{4}$$

where, Gk is Generation of turbulent kinetic energy due to mean velocity gradients, Gb is generation of kinetic energy due to buoyancy, Ym is contribution of fluctuating dilation in compressible turbulence, C1ϵ, C2ϵ, C3ϵ are experimentally determined constants, σt is turbulent viscosity, σk is Prandtl number as a function of k and σϵ is Prandtl number as a function of ϵ. First, the energy variable conditions and determines the turbulence variable, like the K-epsilon model [9,11].

2.2.3. Discrete phase model. Based on this model, the data necessary to simulate the particles are established, contemplating the fluid temperature, sedimentation velocity and its diameter [11], therefore, according to what was proposed by Newton, the rotational and translational particle movements are described in Equation (5) and Equation (6).

$$m\frac{dv}{dt} = mg + \sum F_c + F_d + F_b, \tag{5}$$

$$I\frac{dv}{dt} = \sum T_c, \tag{6}$$

where m is solid mass, v is translational velocity, I is inertia tensor, w is rotational velocity, Fc is contact force, Fd is drag force, Fb is buoyancy, g is acceleration of gravity and Tc is torque contact [12]. Furthermore, the contact force has a normal and tangential component, which is represented in Equation (7).

$$F_c = F_{c,n} + F_{c,T}, \tag{7}$$

where $F_{c,n}$ is normal contact force and $F_{c,T}$ is tangential contact force, consequently the contact torque is equivalent to the tangential torque, as shown in Equation (8).

$$T_c = T_T, \tag{8}$$

where T_c is contact torque and T_T is tangential torque; from what Cundall and Strack (1979) postulate, the normal and tangential contact forces act on the particles, generating the movement through the fluid. Now, to evaluate the contact forces, the model can be linearized, using a spring-damper type model [12] making use of different coefficients to studied particle associated with friction, damping and stiffness, thus obtaining Equation (9) and Equation (10).

$$F_{c,n} = -k_n\delta_n - n_n v_n, \tag{9}$$

$$F_{c,t} = -k_t\delta_t - n_t v_t, \tag{10}$$

where kn is the stiffness, kt is the tangential stiffness, gn is the normal damping coefficient, gt is the tangential damping coefficient, δn is the displacements between particles, δt is in the normal and tangential directions, vn is normal velocity and vt is tangential velocity. The particles sliding is related to Coulomb friction model [13] which is defined as follows (see Equation (11)). Therefore, by replacing in Equation (8), we obtain Equation (12).

$$F_{c,t} = -\frac{f_s|F_{c,n}|\delta_t}{|\delta_t|}, \tag{11}$$

$$T_t = r_n * F_{c,t}, \tag{12}$$

where r, is the normal vector, n, is the radius of the particles.

2.2.4. Fluid motion model. The finite volume method is used to carry out the analysis in addition; the Navier-Stokes equations are used to describe the water flow [14]. The corresponding Equation (13) and Equation (14) are shown below.

$$\frac{\partial}{\partial t}(\varepsilon p_f) + \frac{\partial}{\partial x_f}(\varepsilon p_f u_j) = 0, \tag{13}$$

$$\frac{\partial}{\partial t}(\varepsilon p_f u_i) + \frac{\partial}{\partial x_i}(\varepsilon p_f u_i u_j) = -\frac{\partial p}{\partial x_i} + \frac{\partial}{\partial x_j}\left[\varepsilon\mu\left(\frac{\partial u_i}{\partial x_j} + \frac{\partial u_j}{\partial x_i}\right)\right] F_s, \quad (14)$$

where pf is the density of the fluid, u is the velocity of the water, p is the pressure of the water, μ is the viscosity of the water, x is the denotation of coordinates and the interaction between the forces of buoyancy and Fs is drag [13,14], ε is defined as the specific porosity in the calculation of an empty cell [14], as shown in Equation (15).

$$\varepsilon = 1 - \sum_{i=1}^{N} \frac{V_{p,i}}{V_{cell}}, \quad (15)$$

where N is the total number of particles present in a specific cell of the CFD, Vp,i is the volume of particle and Vcell is the volume of the cell in the CFD, [15]. In this way, the model for the movement of the flow is proposed, contemplating the inherent physical aspects and the guarantee of the dragging of the particles [16].

2.3. Processing (simulation conditions)
In this phase of the simulation development, the solution of the previously proposed mathematical model is carried out. In the first place, the precision is assigned to the solution, to be obtained, in this case, the simple solution monitor is activated, given the model low complexity, and an average solution of 100 iterations is also executed [17], at the end, the data obtained during the sizing and 3D model phase are entered for the particles (velocity and diameter) [18].

2.4. Post-processing
The results are organized in tables, and the visualization of the model is made from a deformed geometry, in this way the flow lines, the particles behavior (sand) and other relevant data can be clearly visualized [19].

3. Results and discussions
The results obtained from the simulation, present several aspects and emergent physical phenomena within each proposed unit, ST, with the hydraulic simulation it is possible to visualize and analyze them to a great extent, however, to determine the highest efficiency and comply with the objective of study, the analysis is limited to velocity and behavior of the sand.

3.1. Flow velocity
The flow velocity inside each ST has been affected by different factors associated with the structure, its geometry, the type of screens and their order; velocity flow is a fundamental aspect in each ST since it guarantees the drag of the solids and the constant flow inside the ST [1].

Figure 1 shows the flow behavior inside each proposed unit, ST; how in the STs the velocity presents variations, due to the flow that occurs in a chaotic and turbulent way, these phenomena also induce the potential energy gain, in the flow and the appearance of vorticity, which in turn produces dead zones in the ST, (see Figure 1(a), Figure 1(c), Figure 1(d), Figure 1(e), and Figure 1(f)) [2], these high and irregular velocities make the flow remain for a short time in the ST and do not guarantee a good settling of the solids. Figure 1(b) shows how the flow passes in order due to the proposed diffusers, so there will be no physical phenomenon that affects hydraulic operation [3].

Figure 2 shows how effectively the ST with double perforated screen presents a slight variation in velocity, while the ST without perforated screens present the most abrupt velocity variations [4]; however, the ST that have a single perforated screen also varies significantly, in contrast to the unit that has two perforated screens [5].

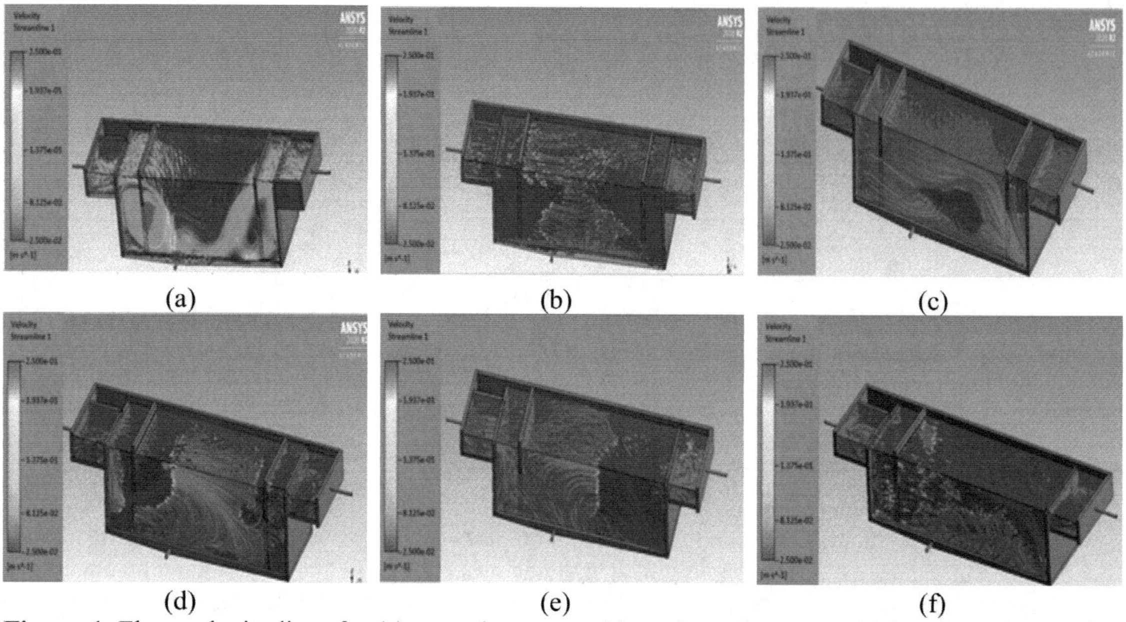

Figure 1. Flow velocity lines for (a) smooth screens, (b) perforated screens, (c) interleaved smooth-perforated screens, (d) Interleaved perforated-smooth screens, (e) ST of a single smooth screen, and (f) ST of a single perforated screen.

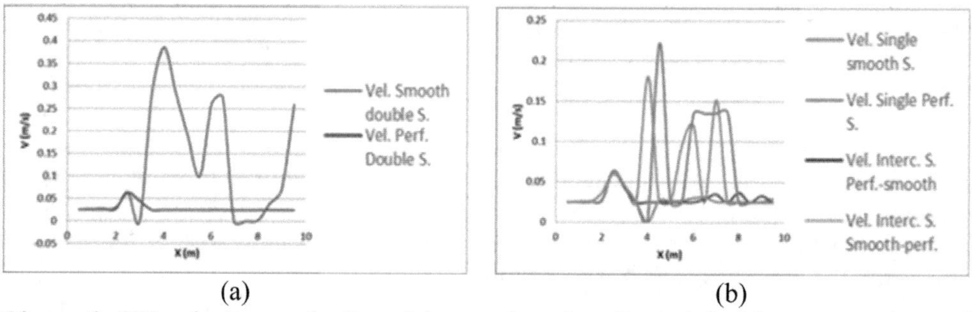

Figure 2. XY velocity graphs from (a) smooth and perforated double screen units, and (b) single and interleaved screen units.

3.2. Behavior of the sands inside each unit

The sands behavior and their transit to the interior of each ST is affected by the emergent physical phenomena and the hydraulic operation of the same, and this in efficiency turn conditions of the STs and their ability to settle solids [6,7]. It can be seen in the Figure 3 that the only ST that presents an ordered flow from the proposed diffusers is the one with a double perforated screen, on the contrary, the other ST present within themselves, turbulent flows, vorticity and dead zones, this influences in a way negative in the ST hydraulic operation since it generates solids resuspension, when they are trapped in the vortex and not find an orderly flow, they tend to go back to the surface, this is evidenced by the efficiency, which does not exceed 50%, while in the ST double perforated screen it is approximately 78% [8].

3.3. Stability of each model from courant number

The courant number is used to determine the model stability; the courant number guarantees stability when your result is close to one [9,10]. Table 1 shows the results obtained for each model as a function of the length (x) and the courant number.

From the values obtained, the analyzed models' stability is evidenced, since the courant number in most of them remains above 0.5 [19], however, there are small unstable areas in the models with double interleaved screen, in which there is a Courant number lower than 0.5, however, its influence on the general stability is minimum [19].

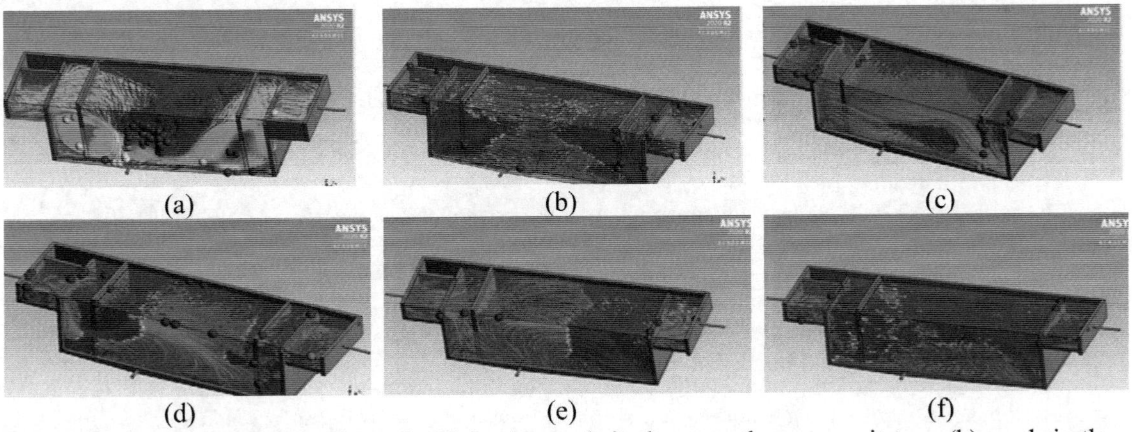

Figure 3. Sands behavior inside each ST for (a) sands in the smooth screen grit trap, (b) sands in the perforated screen grit trap, (c) ST of interleaved smooth-perforated screens, (d) ST of Interleaved perforated-smooth screens, (e) ST of a single smooth screen, and (f) ST of a single perforated screen.

Table 1. Courant number for each unit, based on length (x).

X (m)	# Courant (Smooth double screen)	# Courant (Perforated double screen)	# Courant (Smooth perforated screen)	# Courant (Perforated smooth screen)	# Courant (Single screen smooth)	# Courant (Single perforated screen)
1	0.99820	0.99931	0.93268	0.95241	0.88631	0.91477
2	0.97892	0.99927	0.92007	0.95009	0.90845	0.93301
3	0.95207	0.99910	0.91588	0.92144	0.90001	0.94660
4	0.94071	0.99475	0.91006	0.92004	0.92471	0.92401
5	0.93006	0.99108	0.88375	0.90407	0.77325	0.86247
6	0.90010	0.98375	0.68402	0.84152	0.60028	0.70007
7	0.84063	0.93018	0.62379	0.72560	0.59325	0.62989
8	0.80488	0.97266	0.38541	0.63848	0.84368	0.89224
9	0.89177	0.99117	0.24901	0.49005	0.95884	0.96001
10	0.97114	0.99901	0.56911	0.40792	0.88211	0.90106

4. Conclusions

The importance of this study is based on correcting traditional dimensioning through computational fluid dynamics model, to avoid low efficiencies in projects in which this type of structures is implemented since, having a ST operating with low efficiency, there will be greater turbidity, which implies that a higher coagulant concentration is used in the treatment plant, this represents a high cost in the process and can generate a deficit in the companies in charge of carrying out the treatment. The ST that presented the best efficiency was the double perforated screen (78%), the other ST proposed did not even exceed 50% efficiency. Based on what was obtained, it is recommended to first place at least two consecutive perforated screens in the ST, in order to obtain high removal efficiencies and better hydraulic performance, it is also suggested that for future studies the model be contrasted, with a model real, which provides more accurate and adjusted data.

References

[1] Alghurabi A, Mohyaldinna M, Jufara S, Younis S, Abduljabbara A, Azuwana M 2020 CFD numerical simulation of standalone sand screen erosion due to gas-sand flow *Journal of Natural Gas Science and Engineering* **10(103)** 103

[2] Cundall, P A, Strack, O D L 1979 A discrete numerical model for granular assemblies *Geotechnique* **29** 47

[3] Chang S, Wang X 2003 Multi-dimensional courant number insensitive Euler solvers for applications involving highly nonuniform *39th AIAA/ASME/SAE/ASEE Joint Propulsion Conference and Exhibit* (Huntsville: American Institute of Aeronautics and Astronautics)

[4] Zhou L, Zhao Y 2020 CFD-DEM simulation of fluidized bed with an immersed tube using a coarse-grain model *Chemical Engineering Science Journal* **231** 20

[5] Maddahian R, Asadi M, Farhanieh B 2012 Numerical investigation of the velocity field and separation efficiency of deoiling hydrocyclones *Pet. Sci.* **9** 511

[6] Batista D 2019 Mesh-independent streamline tracing *Journal of Computational Physics* **401** 108

[7] Cajias A, Villablanca A 2014 Diseño de un desarenador para el pre filtrado de agua de riego *Informativo INIA Ururi.* **98** 1

[8] Ratish B, Naidu K 1993 Hydrodynamics of bubble column reactors at high gas velocity: experiments and computational fluid dynamics (CFD) simulations *Industrial & Engineering Chemistry Research* **46(25)** 8431

[9] Parsi M, Agrawal M, Srinivasan V, Vieira R E, Torres C F, McLaury B S, Shirazi S A 2015 CFD simulation of sand particle erosion in gas-dominant multiphase flow *Journal of Natural Gas Science and Engineering* **27(2)** 706

[10] Shirzadi M, Parham Mirzaei P, Naghashzadegan M 2017 Improvement of k-epsilon turbulence model for CFD simulation of atmospheric boundary layer around a high-rise building using stochastic optimization and Monte Carlo Sampling Technique *Journal of Wind Engineering and Industrial Aerodynamics* **171** 366

[11] Saidi M, Maddahian R, Farhanieh B, Afshin H 2012 Modeling of flow field and separation efficiency of a deoiling hydro cyclone using large eddy simulation *International Journal of Mineral Processing* **112(113)** 84

[12] Wang J, Guo F, Zhao Q, Wu M 2013 Numerical simulation of flow field in underground cyclone desander with different entrances *Advanced Materials Research* **655(657)** 470

[13] Gorobets A, Tarabara V, 2020 Separation performance of desanding and deoiling hydrocyclones treating three-phase feeds: effect of oil-particle aggregates *Separation and Purification Technology Journal* **237** 10

[14] Liang X, He L, Luo X, Li Q, You Y, Xu Y 2021 Experimental study on sand particles accumulation, migration and separation efficiency in slug catcher *Chinese Journal of Chemical Engineering* **32** 134

[15] Siaw K, Yung J, 2019 CFD study of sand erosion in pipeline *Journal of Petroleum Science and Engineering* **176** 269

[16] Awad A M, Hussein I A, Nasser M S, Karami H, Ahmed R, 2020 CFD modeling of particle settling in drilling fluids: impact of fluid rheology and particle characteristics *Journal of Petroleum Science and Engineering* **199** 108

[17] Fukada T, Takeuchi S, Kajishima T 2014 Effects of curvature and vorticity in rotating flows on hydrodynamic forces acting on a sphere *International Journal of Multiphase Flow* **58** 292

[18] Jeong W, Seong J 2014 Comparison of effects on technical variances of computational fluid dynamics (CFD) software based on finite element and finite volume methods *International Journal of Mechanical Sciences* **78** 19

[19] De Plaza J, Pulgarín., Ruge J, Rojas J 2021 Hydraulic modeling of combined sewers overflow integrating the results of the SWMM and CFX models *Rev. Int. Métodos Numér. Cálc. Diseño Ing.* **37(1)** 14

Bottom rack intake improvement as a fluid physics application through a computational fluid dynamics model

H E Calderón[1], L M Rada[1], and J S De Plaza[1]
[1] Facultad del Medio Ambiente y Recursos Naturales, Universidad Distrital Francisco José de Caldas, Bogotá, Colombia

E-mail: hecalderong@correo.udistrital.edu.co, sjsebastiand@correo.udistrital.edu.co

Abstract. This research focuses on improving the hydraulic behavior of a traditionally design bottom rack intake, from variations in roughness parameters, free height, and the inclusion of chamfers, establishing a contribution to the contrast between classical physics and the physics that takes over the partial resolution of the Navier-Stokes equations. To make possible the structure in OpenFOAM, it is necessary to use the geometric tool Salome-Meca, as well as a meshing tool (snappyHexMesh), and the InterFOAM solver in the processing stage. In the same way, through the turbulence model (K-E) local effects are evidenced in the Fluid-Structure interaction, as well as the identification of events and the development of the phenomenon of vorticity. The results show the improvement presented in some areas of the structure from the stabilization of the water flow through of the fluid-structure interaction change, the modification of the geometry and roughness, minimizing the presence of vertical vortices, cavitation, and surrounding areas. This allows us to conclude that traditional hydraulic do not consider the real physical flow behavior within the structure and neither the subsequent phenomena that develop, establishing as a starting point the need to rethink the design of the bottom rack intakes.

1. Introduction
A bottom rack intake (BRI) has the function of capturing the water in high mountain rivers. This, through a rack with bars located in the flow direction, which is responsible for diverting the water to the conduction pipeline. To prevent damage to the structure or drinking water treatment plant (DWTP), it is possible to simulate the future hydrodynamic flow behavior over structures, with the purpose of predicting phenomena presented by changes in the system variables. This, using computational fluid dynamics (CFD), from the OpenFOAM open-source software, which uses Reynolds averaged Navier-Stokes equations (RANS) [1], and provides a three-dimensional solution to practical problems.
This software also allows observing the system air-water and fluid-structure interaction, using finite volumes method (VOF), and the InterFOAM solver (which has proven to be reliable for the simulation of free surface vortices in an intake) [2].

Various studies have shown hydraulic failures in dimensioned under traditional parameters (BRI), as in the case of [3], which shows the vortex development due to the momentum generated when the flow impacts on the water collection channel surface. Likewise, in that area, according to [4], there are unstable lines flow phenomena that triggers eddies by the air capture. Similarly, in [5], it is stated that, if although CFD is used to improve the knowledge of hydraulic phenomena, it is not possible to analyze effects which occur in the bars, due to their multiplicity. Furthermore, in [6], it is claimed that the surrounding velocity vectors are the effect of the square geometry of the vertex, both in channel and in water collection chamber, which leads to the drain and excesses pipeline. However, the effects generated by the flow dynamics in the main walls area, as which are appeared in the water collection channel and chamber, have not been subjected to an exhaustive study. That is why, in this research, it is proposed to generate an alternative of hydraulic improvement in a BRI, from the Hydraulics deficiencies identification of the traditional design and structure simulation in CFD as an earlier support of traditional engineering projects that consider the physical flow behavior inside the structures.

2. Methodology and materials

To obtain results, two phases are planted that, carried out, analyze the flow behavior through the structure. The first objective focused to design the BRI under the classical hydraulics parameters, in order to establish its geometry. Then, it is proposed to carry out the 3D structure modeling in three instances; preprocessing, to create the solid geometry, processing meshing and computational modeling through OpenFOAM, and post-processing to generate the graph results.

2.1. Traditional design phase

The BRI, composed of the intake rack, channel, collection chamber, and water excess weir, is designed based on classical hydraulics parameters, in relation to sizing, flow velocity ranges, and free heights recommendations. On the other hand, the design flow is 0.026 m^3/s for structures dimensioning.

2.2. OpenFOAM modeling phase

The computational modeling is carried out through OpenFoam 8.0, and is executed from the following steps: solver, turbulence model, vorticity model and model initial conditions.

2.2.1. InterFOAM solver. Due to the fluids characteristics to be modeled (incompressible, isothermal, and immiscible), the interaction model between phases is carried out with InterFOAM solver, which solves the three-dimensional equations for the water-air interface [7], using the fluid volume method (VOF); this, assuming a constant water and air density [7].

Equation (1) represents the momentum conservation [2], where u is the velocity, ρ is the fluid density, g is the gravity acceleration, and p is the pressure. On the other hand, x_i and x_j represent a fixed Cartesian system coordinate, and T_{ij} and T_{tij} are the viscous and turbulent stresses. Similarly, $f_{\sigma i}$, is the surface tension.

$$\frac{\partial(\rho u_i)}{\partial t} + \frac{\partial}{\partial x_j}\left(\rho u_j u_i\right) = \frac{\partial p}{\partial x_i} + \frac{\partial}{\partial x_j}\left(T_{ij} T_{tij}\right) + \rho g_i + f_{\sigma i}, \tag{1}$$

$$\alpha = \left| \begin{matrix} 0 \\ 0 < \alpha < 1 \\ 1 \end{matrix} \right| \begin{matrix} \text{Phase 2} \\ \text{Interface,} \\ \text{Phase 2} \end{matrix} \tag{2}$$

where α is the unit of water density (ρ: 1), and 0 represents the density of air (ρ: 2). At the interface between the two fluids, α tends to vary between 1 and 0. So, in Equation (2), if the calculation of α is 1, the density of the liquid in that part of the mesh is the density of water; if α, on the contrary, is zero, it takes the density of the air in each cell, α being a scalar quantity, which describes the volume of a cell for a certain phase [8].

2.2.2. Turbulence model. A K-epsilon turbulence model is implemented, as it contains the governing equations developed for incompressible turbulent flows, with high Reynolds numbers.

2.2.3. Turbulent kinetic energy, k. The k-ε model contains two transport equations that consider the turbulent flow effect on the structure [9] and its dissipation rate. The first of these determines the Convection Effect, and the second, the Turbulent Energy Dissipation Rate (see Equation (3) and Equation (4)).

$$\frac{\partial}{\partial t}(\alpha\rho\epsilon) + \nabla\cdot(\alpha\rho u\epsilon) - \nabla 2(\alpha\rho D_\epsilon \epsilon) = C_1 \alpha\rho G \frac{\epsilon}{k} - \left(\left(\frac{2}{3}C_1 - C_3\right)\alpha\rho\nabla\cdot u\epsilon\right) - \left(C_2 \alpha\rho \frac{\epsilon}{k}\epsilon\right), \tag{3}$$

$$\frac{\partial}{\partial t}(\alpha\rho k) + \nabla\cdot(\alpha\rho u k) - \nabla^2(\alpha\rho D_k k) = \alpha\rho G - \left(\frac{2}{3}\alpha\rho\nabla\cdot u k\right) - \left(\alpha\rho\frac{\epsilon}{k}k\right), \tag{4}$$

where α is the phase fraction, ρ is fluid density (kg/m^3), ε is the turbulent kinetic energy dissipation rate (m^2/s^3), u is the flow velocity (m/s), D_ϵ is the effective diffusivity for ε, G is the turbulent kinetic energy production rate due to the anisotropic part of the Reynolds stress tensor (m^2/s^3), k is the turbulent kinetic energy production rate due to the anisotropic part of the Reynolds stress tensor (m^2/s^3), C_1 and C_2 are model coefficient (1.44), (1.92) and D_k is the fast distortion theory compression term (0.0).

In the same way, wall and roughness functions are adapted according to the type of material and the model limits. Equation (5), Δt approach, represents the wave velocity through the mesh (must be less than unity); Equation (6), on the other hand, represents the turbulent energy dissipation rate in m^2/s^2. Lastly, Equation (7), turbulent kinetic energy is characterized by velocity fluctuations in a highly turbulent flow [5].

$$\Delta t = C\frac{\Delta x}{u}, \tag{5}$$

$$\varepsilon = \frac{C\mu^{0.75}k^{1.5}}{L}, \tag{6}$$

$$k = \frac{3}{2}\left(I(u)\right)^2, \tag{7}$$

where C is the Courant number, Δx is the minimum mesh size (m), L is the turbulent scale length (m), and I is the turbulence intensity (0.07 %) [9].

3. Results and discussion

In this section, it is shown the results obtain through of the model analysis carried out under conventional hydraulics. Then, the comparison of it with the previously improvement described.

3.1. Model Analysis under conventional hydraulics

The main walls sizing, based on the conventional design (at 45°), originates a fluid structure interaction (FSI). This generates incident shocks, whose convergence occurs in the longitudinal focus of the intake throat [10,11]. Consequently, the turbulence intensity at this point increases due to the fluctuation imposed by the wall on the velocity vectors, producing a variation in flow depth (with a maximum upstream of 0.202 m and a minimum downstream of 0.07 m). At the same time, by gradually reducing the width of the channel, a transition is generated from a sub-critical regime upstream of the flow to a super-critical one downstream, affecting the Froude number increase (from 0.071 to 1.128); this happens, as the river flow is channeled through a smaller area, which implies downstream flow resistance. Consequently, the hydraulic jump formation effect in the throat is obtained.

On the other hand, it is observed that the effect on the bars that make up the grid, under a super-critical flow regime, produces an undulating adhesion effect and flow drag when it passes over it. This induces the "leaf jet" effect, in which the water tends to remain bound on the bars in the rack (Figure 1), which coincides with the results obtained in [12]. Also, considering a tilt ratio, the fall of the flow is greatly contributed, producing greater turbulence and increased pressure drop over them (from 0.086 m to 0.01 m). Concerning the flow behavior over the racks, a progressive decrease in kinetic energy and the join is evidenced, which generates a hydrostatic pressure gradient along the bars, from 1663.62 Pa, at the beginning of the rack, to 589.028 Pa at the end of this [13]. It is also observed that, at the beginning of the bars, there is a sub-critical condition (Froude 0.1) with a transition to super-critical regime at the end of the bar (Froude 7.36).

At the same time, a dissipation effect is generated at the end of the bar, as evidenced in [4], since the oscillating interaction between the velocity and the tie falls unstably on the bars. This induces a spatially varied flow (SVF) with decreasing discharge, as shown in the experimental results of [12,13] (see Figure 2), where such a flow condition is evidenced in the rack.

Figure 1. Leaf jet.

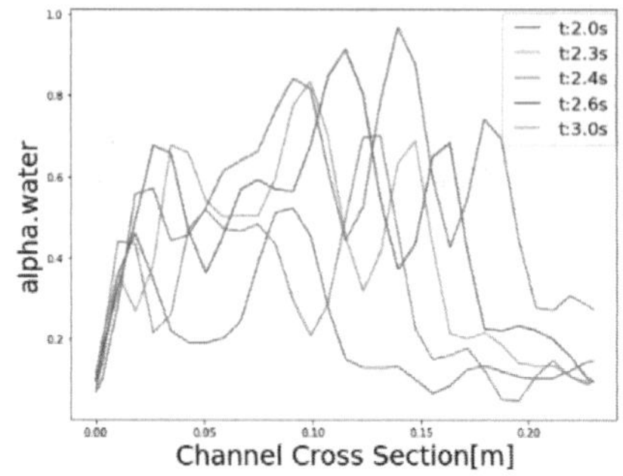

Figure 2. Spatially varied flow.

Since the shock current wave, it supposed the recirculation in the wall of the channel (because the angular momentum conservation). A water-air mixture is produced (Figure 3), which spreads the flow aeration over the water collection channel walls, this being relevant, since it benefits the water reoxygenation in its treatment process, up to a value of 6485 mg/L [14]. In the same way, since the discharges are variable throughout the channel length, there is an increase in the join (of 0.015 m on the opposite channel wall and 0.0931 m on the adjacent wall to the flow direction) and upward flow, whose lines are considerably curved and sloping, before delivering to the collection chamber, resulting in a flow rotational discharge. However, intense turbulence is observed, as well as a channel domain interface, which generates small waves, product of the fluctuations caused by the vertical vortex presence [15].

In addition, it is evident that the receiving shock wave wall increases the pressure, generating less sensitivity to the phenomenon already described. However, in the center and adjacent wall, tends to present both vacuum pressure (Figure 4) (on average 1292 Pa and 492 Pa, respectively), as pressures below the water vapor pressure at 15 ° (1657.27 Pa). This supposes that the water nappe hits the channel opposite wall, passing over the channel floor, and subsequently collides with the adjacent wall. Thus, it begins to adhere to the surface, as the vortex develops. This tends to detach, generating the effect of suction (Figure 4), which in turn implies the presence of pressures below atmospheric pressure, promoting the cavitation risk in the area [16].

Figure 3. Channel vorticity.

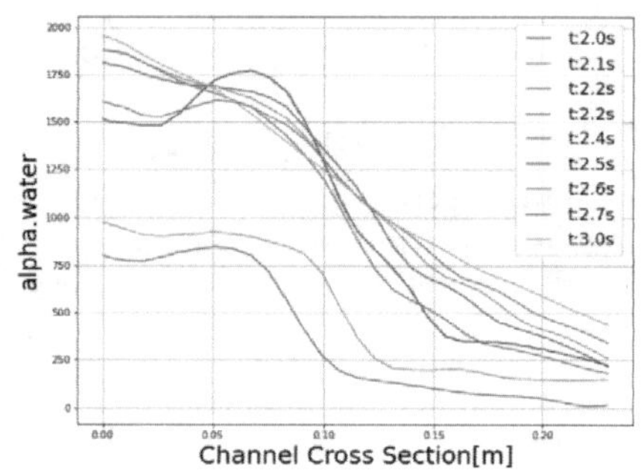

Figure 4. Pressure on adjacent wall.

Given that the cavitation index indicates the cavitation existence in the fluid [17], Table 1 shows the presence of this phenomenon in the channel floor and in the water collection channel adjacent wall. In addition, the accentuated layer turbulent discharge (water collection channel overhang), adopts a flow rotational pattern lines at the wall corners, generating a 0.078 m pressure drop, low kinetic energy and, on the contrary, higher potential energy.

Regarding the layer effect on the bottom of the water collection chamber, it is observed that the height of fall on the chamber increases the jet potential energy, which implies an increase in flow velocity (from 2m/s to 6 m/s) downstream of the weir (Figure 5). Likewise, it promotes the dead zones formation, due to the angular velocity of 2.86 Hz, present in the chamber corners (see Figure 6).

Journal of Physics: Conference Series **2118** (2021) 012003 doi:10.1088/1742-6596/2118/1/012003

Table 1. Cavitation index.

	Velocity (m/s)	Pressure (Pa)	Cavitation index (σ)	
Channel opposite wall	0.15	1678.924	1.86	No cavitation
Channel floor	0.34	1292.435	-6.73	Cavitation
Channel adjacent wall floor	0.29	491.360	-27.85	Cavitation

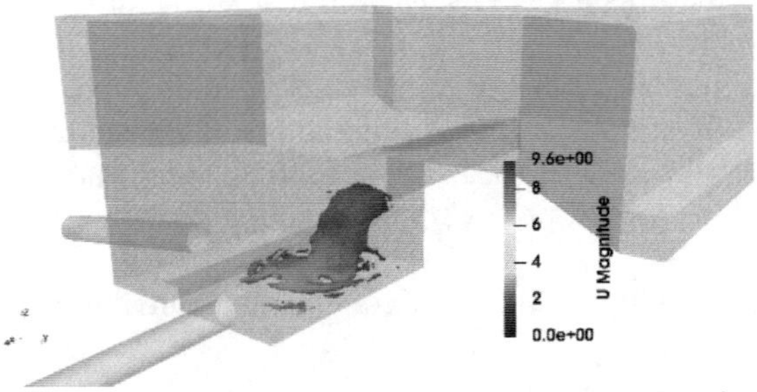

Figure 5. Unloading the water nappe in the collection chamber.

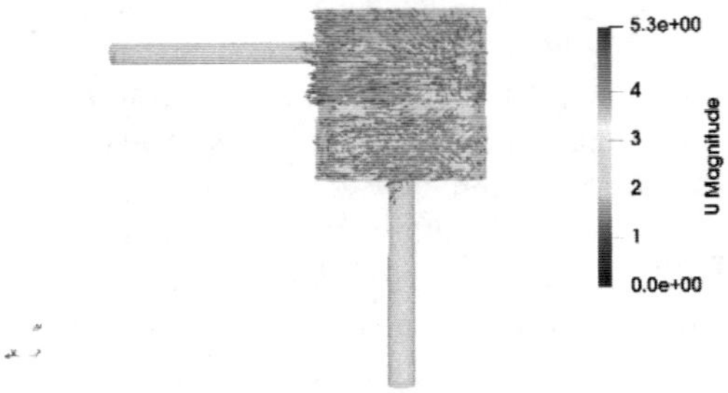

Figure 6. Flow lines in the collection chamber.

3.2. Improvement to the conventional design from the computational fluid dynamics model

The proposed geometric alternative (whose characteristics are summarized in Table 2), makes possible the flow field appeasement due to the implementation of chamfers in the channel corners and chamber water collection (see Figure 7). This, in turn, favors the uniform distribution of velocities from the channel to the chamber, which implies the turbulence reduction and hydraulic improvement (Figure 8).

In the same way, from the modifications made, the dead zones reduction in the channel is evidenced, which entails the hydraulic load reduction, with respect to the one previously presented towards the of the collection channel corners from 0.167 m to 0.075m, as it is possible to see in Figure 8.

Equally, about the change in roughness (from 0.03 to 2mm), it is observed that the flow lines at the bottom are disturbed due to the rough walls at the bottom and concrete surfaces in the structure. Indeed, a behavior isotropic convergent flow in interaction with the rough wall [18]. In addition, a greater velocity gradient is obtained as a result, due to the velocities fluctuation generated by the surface rebounding, as well as the flow energy dissipation due to the roughness increase.

Table 2. Improvement characteristics summarize.

		Value (m)
Slope (m/m)		0.05
Chamfer radius	Collection channel	0.25
	Collection chamber	0.10
Free boundary (additional to the recommended minimum)		0.17

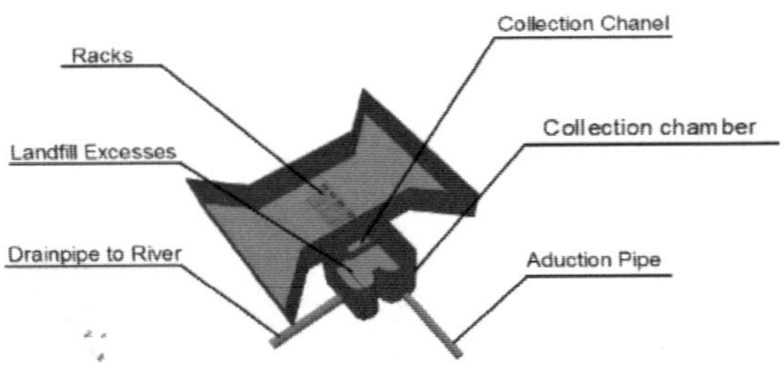

Figure 7. Unloading the water nappe in the collection chamber.

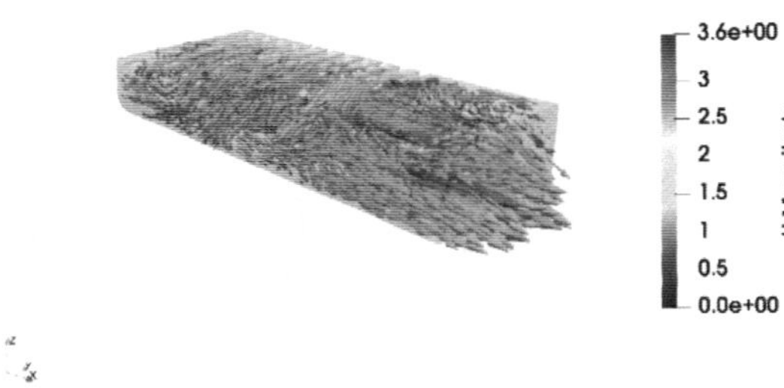

Figure 8. Flow lines in the collection chamber.

4. Conclusions

The geometric proposal allowed to improve the hydraulic-structural performance in the bottom rack intake, minimizing the eddies presence from the incorporation of chamfers, and the consideration of an additional free edge to that considered in traditional hydraulics. This, if applied, would prevent damage to the bottom rack intake concrete structure. The research carried out allows us to infer that the traditional hydraulics for the bottom rack intakes design does not consider the fluid-structure interaction, ignoring the vorticity effects and recircundant zones that this generates, as well as the spatially varied flow presented in the collection channel.

Likewise, it is suggested to apply a Lagrangian model to consider the sands interaction with the structure, and thus demonstrate the effect of viscous forces on energy dissipation. This, to adjust the phenomenon to the physical model. Also, it is recommended to run this under a direct numerical simulation or large eddy simulations model type, which would provide a closer approximation to the turbulent effect of flow on the structure.

Finally, it is recommended to add energy dissipation structures in the water collection channel to reduce flow energy to the chamber collection.

References

[1] Eichhorn M, Doujak E, Waldner L. 2016, Investigation of the fluid-structure interaction of a high head Francis turbine using OpenFOAM and Code_Aster *IOP Conf. Series: Earth and Env.Sci.* **49(1)** 1

[2] Domfeh M, Gyamfi S, Amo-Boateng M, Andoh R, Ofosu E,Tabor G 2020 Numerical simulation of an air-core vortex and its suppression at an intake using OpenFOAM *Fluids* **5(4)** 221

[3] Brunella S, Hager W, Minor H 2003 Hydraulics of bottom rack intake *J. of Hydraulic Engineering* **129** 2

[4] Tokyay N, Velioglu D, Dincer 2015 A numerical and experimental study on the characteristics of hydraulic jumps on rough beds *Proceedings of the 36th IAHR World Congress* (Hague: International Association for Hydro-Environment Engineering and Research)

[5] Carrillo J, Garcia J, Castillo L 2018 Experimental and numerical modelling of bottom intake racks with circular bars *Water* **10** 5

[6] Marghzar S, Montazerin N, Rahimzadeh H 2003 Flow field, turbulence and critical condition at a horizontal water intake *Journal of Power and Energy* **217** 53

[7] Ferrer P M, Causon D M, Qian L, Mingham C,Ma Z H 2016 A multi-region coupling scheme for compressible and incompressible flow solvers for two-phase flow in a numerical wave tank *Computers & Fluids* **125** 116

[8] Spitzenberger A, Neumann S, Heinrich M, Schwarze R 2020 Particle detection in VOF-simulations with OpenFOAM *SoftwareX*, **11** 100382

[9] Tsan-Hsing S, Nan-Suey L, Kuo-Huey C 1998 A non-linear k-epsilon model for turbulent shear flows *34th Joint Propulsion Conference and Exhibit* (Cleveland: American Institute of Aeronautics and Astronautics)

[10] Defina A, Pietro D 2010 Open channel flow through a linear contraction *Physics of Fluids* **22** 036602

[11] Tosi L P, Colonius T 2019 Modeling and simulation of a fluttering cantilever in channel flow *Journal of Fluids and Structures* **89** 174

[12] Chan S N, Wong K C C, Lee H W J 2018 Hydraulics of air–water flow in a supercritical bottom rack intake *Journal of Hydro-Environment Research* **21** 60

[13] Castillo L G, García J T, Carillo J M 2017 Influence of rack slope and approaching conditions in bottom intake systems *Water* **9(1)** 65:1

[14] Tsai C H, Yeh C P, Lin C H, Yang R J, Lin C H, Lin C H, Fu L M 2012 Formation of recirculation zones in a sudden expansion microchannel with a rectangular block structure over a wide Reynolds number rang *Microfluid Nanofluid* **12** 213

[15] Chen Y L, Wu C, Ye M, Ju X M 2007 Hydraulic characteristics of vertical vortex at hydraulic *Journal of Hydrodynamics, Ser. B* **19(2)** 143

[16] Turalina D, Yembergenova D, Alibayeva K 2015 The experimental study of the features of water flowing through a sharpcrested weir in channel *EPJ Web of Conferences* ed. Dančová P, Vít T (Český Krumlov: EDP Sciences)

[17] Hosseinpour Shafaghi A, Rokhsar Talabazar F, Zuvin M, Talebian Gevari M, Villanueva L G, Ghorbani M, Kozar A 2021 On cavitation inception and cavitating flow patterns in a multi-orifice microfluidic device with a functional surface *Physics of fluids* **33(3)** 032005

[18] Orlandi P, Leonardi S 2006 DNS of turbulent channel flows with two- and three-dimensional roughness *Journal of Turbulence* **7** N73:1

Differential microwave sensor based on microstrip lines loaded with a split-ring resonator for dielectric characterization of materials

J Zapata-Londoño[1], F Umaña-Idárraga[1], J Morales-Guerra[1], S Arias-Gómez[1], C Valencia-Balvin[1], and E Reyes-Vera[1]
[1] Departamento de Electrónica y Telecomunicaciones, Instituto Tecnológico Metropolitano, Medellín, Colombia

E-mail: erickreyes@itm.edu.co

Abstract. In this work, we propose a microwave sensor that allows the characterization of dielectric materials based on a differential configuration. A microstrip permittivity sensor of the surrounding material is proposed using a split ring-resonator to measure differentially. The geometry was optimized and was numerically analyzed using CST STUDIO. The numerical analysis of the metamaterial unit cells is carried out first, to determine the operating band. After that, the metamaterial cell was employed to design the differential microstrip permittivity sensor. The obtained results reveal that the proposed sensor has the capability to characterize different materials whose relative dielectric permittivity's are in the range of 9.8 to 80 with great performance. The device has a total size of 86 mm × 60 mm and operates around 3 GHz. In this band, the sensor reaches a sensibility of 2.89 MHz and a Q-factor of 70.15. Thus, this work shows a compact, reusable, label-free, and non-destructive microwave sensing device and paves the way for high accuracy sensing of the dielectric properties of different materials due to its high- Q-factor as well as high sensitivity.

1. Introduction

In recent years, the measurement of the dielectric permittivity of materials has aroused strong interest in many industrial applications due to the fact that the dielectric permittivity of bulk materials can be related to other important parameters such as density, concentration, temperature and humidity [1,2]. Thus, the characterization of this parameter is a powerful way to develop applications or devices to solve problems in medicine, biology, security, food industry, chemical, among others [3–8].

Several microwave methods had been explored to measure the dielectric permittivity of liquids and solids, the main techniques include the free space technique, and the use of resonators, parallel plate capacitors, and microstrip technology [1,2]. However, microwave devices based on microstrip technology emerge as an interesting alternative because they are inexpensive, easy to construct, have simple integration setups with electronic components, and are versatile in comparison with other microwave methods. Likewise, the devices based on the resonant technique offer higher accuracy and high-quality factor (Q-Factor). Moreover, this kind of sensor determines the dielectric properties of the sample under test (SUT) by simply measuring the shift in the resonant frequency, the phase change, and the Q-Factor of the sensor as a function of the dielectric permittivity [5,9,10]. Then, the combination of both characteristics had been explored in the past to achieve high-performance microwave sensing devices. Many kinds of resonant structures such as split-ring resonators (SRRs), step impedance resonators, substrate integrated waveguide (SIW) resonators and others have been employed to date to measure dielectric permittivity [5,11–13]. However, the measurement of this property could be affected by environmental conditions such as humidity or temperature fluctuations, which can induce cross-sensitivity that entails measurement errors.

To avoid these systematic errors due to cross-sensitivity, some interesting alternatives had been proposed. The first one explores the use of materials that are not sensitive to environmental changes, but these materials are expensive in most cases. The second one explores the use of a differential configuration, which compares the stimulus that suffers two identical sensors subject to the same external factors. This alternative could be used in combination with microstrip technology to obtain compact devices that offer the possibility to compensate environmental changes. For example, A. Ebrahimi, *et al.* [14], in 2018, present a microwave sensor using a pair SRRs. In this configuration, the authors load a microstrip line with two identical SRRs on its sides. Thus, a differential permittivity sensor was analyzed and employed to compare a SUT with a reference sample. Simultaneously, Paris-Velez, *et al.* Reported a differential permittivity sensor based on a pair of uncoupled microstrip lines, which are loaded with open complementary split-ring resonators (OCSRRs) [10]. Then, the authors implement this configuration to measure the dielectric changes in mixtures of ethanol and deionized water. Thus, they demonstrate that this kind of configuration is more sensitive and versatile. Other configurations based on Magnetic resonators composed of capacitive and inductive structures [15], Rat-Race couplers [16], and dumbbell-shaped defect ground structures [13] have been recently reported, demonstrating that sensors based on differential configuration are an interesting option when you want to override the effects of the environment.

Therefore, we propose a differential permittivity sensor based on a microstrip coupler loaded with a couple of SRRs, one in each branch. This sensor initially operates at 3.5 GHz. In comparison with other previous studies, the proposed structure has some important advantages. The aim of this work is to demonstrate that the proposed sensor presents a high sensitivity when it is used to measure dielectric changes, and it offers a differential measurement, which can be used to obtain a more stable solution.

2. Methodology and materials

This section details the methodology to analyze the resonator used to verify that it has metamaterial characteristics. Subsequently, the design of the proposed configuration to measure permittivity changes is presented. Finally, the principle of operation of this device is explained to ensure a better understanding of the work.

2.1. Metamaterial unitary cell

Figure 1(a) shows the proposed metamaterial unit cell. This metamaterial cell consists of a unique circular split ring resonator with a gap (g) equal to 2 mm, an internal radio (r) of 4.5 mm, and an outer radio (R) of 5.8 mm. The ring resonators are made of copper lines with a metal thickness of 35 μm. The performance of the proposed unit cell was numerically evaluated using two floquet ports to study its electromagnetic performance, while perfect electric condition (PEC) and perfect magnetic boundary conditions (PMC) were imposed on the metamaterial unit cell, see Figure 1(a). The structure was excited using a uniform plane wave, which is propagated on the z-axis with an open-end boundary condition, visible in Figure 1(a). Likewise, PEC was used in z-direction, a PMC was defined in the XY plane, while the electromagnetic wave was propagated in x-direction. Therefore, the E-field of the incident wave is polarized along x-axis, while the H-field is polarized along the y-axis.

Therefore, the Nicolson Rose Weir (NRW) method was employed to extract and get the permeability and permittivity of the proposed structure from the scattering parameters (S-parameters). This numerical analysis was carried out from 2 GHz to 4 GHz. The obtained results are illustrated in Figure 1(b). The simulations show the real and imaginary parts of the permeability and permittivity of our resonator. Likewise, the results reveal that the proposed SRR presents a negative behavior for the real part of the permeability from 3.2 GHz to 3.49 GHz. In a similar way, the imaginary part of the permittivity has negative values from 3.2 GHz to 3.33 GHz. On the other hand, the real part of the permittivity and the imaginary part of the permeability both remain positive in the studied frequency range. Thus, we can affirm that this structure presents a typical behavior of a single negative (SNG) metamaterial resonator [17–19].

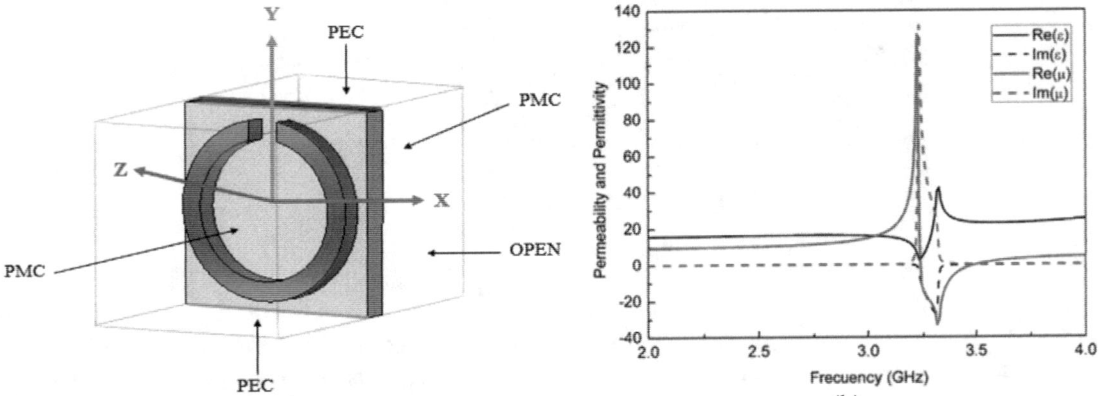

(a) (b)

Figure 1. (a) Schematic of the proposed metamaterial unitary cell, (b) obtained results of the permittivity and permeability of the proposed unitary cell.

2.2. Microwave sensor design

Figure 2 illustrates the proposed microwave sensor, which is based on a differential configuration. In this case, a resistive power divider is employed to split the input power into two identical branches. The lower arm is known as the reference arm, while the upper arm is the sensor arm, and it is used to measure the dielectric permittivity of the SUT. Both arms were designed using a standard microstrip line with an impedance of 100-Ω. Likewise, these microstrip lines are loaded with a circular SRR (the inset of Figure 2 shows the structure with more details). As shown in section 2.1 this metamaterial structure was optimized to resonate at 3 GHz and the resonator was set at a distance of 2 mm from the microstrip line to ensure a good power coupling [20,21]. The SRR structure was selected to design the sensing region of the proposed sensor due that these types of structures help to increase the capacitance of the microstrip circuit, and the *E*-Field in this point is confined. Thus, the gap of this structure allows for a more sensitive area [18]. To model and characterize the performance of this sensor a container with a cylinder shape and diameter of 2 mm was placed on the gap of the SRR.

On the other hand, the numerical analysis of the proposed configuration was carried out using CST microwave studio, which is a software based on the standard finite-difference time-domain (FDTD) method. The substrate material considered in this study was a Rogers 4003C with a relative dielectric permittivity (ε_r) of 3.55, a thickness of 1.52 mm and a tangent of losses of 0.0027. In addition, a thin layer of copper of 35 µm was employed in the construction of the microstrip lines and the ground plane. Table 1 summarizes the most important geometrical parameters of the proposed microwave sensor configuration.

Table 1. Dimensions of the proposed microwave sensor.

Parameter	Value (mm)	Parameter	Value (mm)
W	86.0	d	0.2
L	60.0	s	4.4
lb	54.9	g	2.0
ls	5.2	R	5.8
a	15.4	r	4.5
wa	28.4	hc	4.0
wb	30.8		

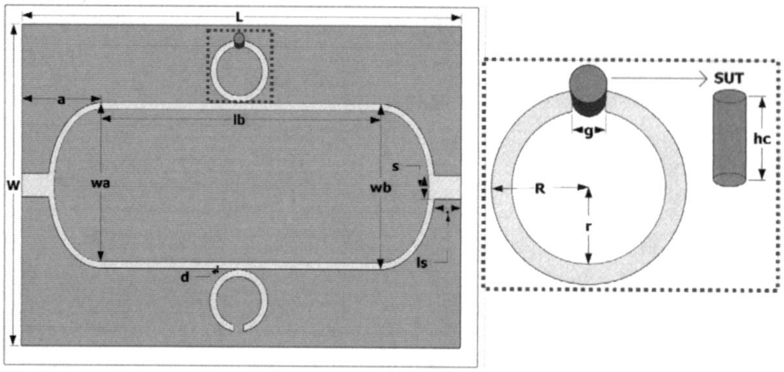

Figure 2. Schematic of the proposed microwave sensor.

2.3. Working principle

The working principle of the proposed microwave sensor is based on mode conversion, which occurs when the sensing branch is loaded with a different material, thus the balanced condition of this configuration is broken. Then, this sensor presents two different situations: the first one is when the configuration is balanced, it occurs when both branches are surrounded by the same material, and the phase difference between both two paths is zero. For this reason, only one resonant peak is detected in the transmission spectrum associated with the natural resonance of the SRR. Conversely, when the sensing branch is in contact with a material and the reference branch is in the presence of a different one, such difference is detected through mode conversion. This mode conversion is induced by the phase difference between the waves that travel through the two paths. In this condition, a second zero in the transmission response of the proposed device is obtained. Then, a variation on the relative permittivity of a SUT could be measured using the difference between the two transmission zeros. This aspect helps obtain a self-referenced sensor configuration. Likewise, these kinds of sensors are immune to external noise due to thermal or humidity changes. Therefore, it is a robust configuration that can provide greater reliability and better stability than other previously reported alternatives.

3. Results and discussions

First, the proposed microwave device illustrated in Figure 2 was simulated using CST STUDIO without a sample to corroborate the initial performance of this sensor device. Figure 3(a) shows the response of the S_{21} parameter between 2.6 GHz and 3.4 GHz. It is clear from this spectrum that the proposed microwave device has a dip at 3 GHz of –16 dB. On the other hand, the spectrum reveals a single resonant peak due that both resonators (reference arm and sensing arm) having the same structure and the surrounding media for both being air (ε_r=1) [12,13]. Therefore, both SRRs resonate at the same frequency. So, it can be said that the device is balanced in this case. In addition, the **E**-Field distribution of the sensor was included at 3 GHz. From Figure 3(b) is easy to corroborate that the **E**-field intensity is higher in the gap region.

As mentioned in section 2.3, any asymmetrical perturbation in the SRR of the sensing arm can induce a mode conversion, so, the transmission spectra will present two dips. Therefore, we can use this working principle to characterize the dielectric permittivity of any SUT. Thus, a simulation process was achieved to study the performance of this microwave sensor when the dielectric relative permittivity (ε_r) values of the SUT are varied. Figure 4(a) shows the obtained results of the magnitude of the S_{21} parameter when the dielectric permittivity of the sample is varied from 9.8 to 80. Now, we can observe the presence of a new resonant peak, whose resonant frequency shifts towards lower frequencies. On the other hand, one peak remains almost constant at 3 GHz for all loaded samples. These results corroborate the capability of this novel structure to give a differential measurement. In fact, this sensor can be self-referenced measurements are always taken with respect to the reference peak. Another advantage of this configuration is that the magnitude of the resonant peak is consistent for all samples, making it easy to implement in real applications.

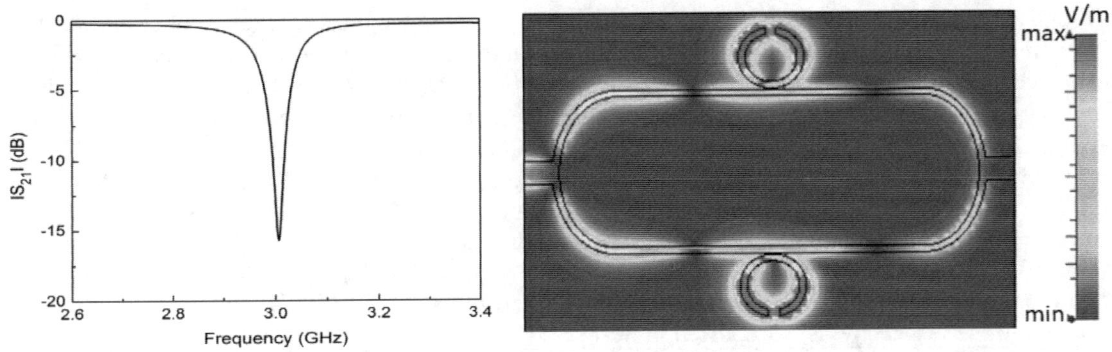

(a) (b)

Figure 3. (a) Frequency response of the proposed differential microwave sensor at the balanced condition, (b) **E**-Field distribution at 3 GHz.

An analysis of the sensitivity and *Q*-Factor of the sensor was carried out using the obtained spectra in Figure 4(a). Figure 4(b) reveals that the proposed differential configuration exhibits high sensitivity. In fact, the resonant frequency change from 2.975 GHz to 2.826 GHz when the dielectric constant varies from 9.8 to 80. Thus, the sensitivity ($\Delta f/\Delta\varepsilon_r$) of this sensor is 2.89 MHz/ε_r. In addition, the exponential adjustment of the obtained results is expected due that these kinds of sensors tend to saturate for high dielectric permittivity values. The quality factor was also evaluated for each sample. Table 2 summarized the obtained results, and evidence that this important parameter remains stable between 64.99 and 70.15. This result tells us again that the shape of the resonance peak is maintained throughout the measurement range.

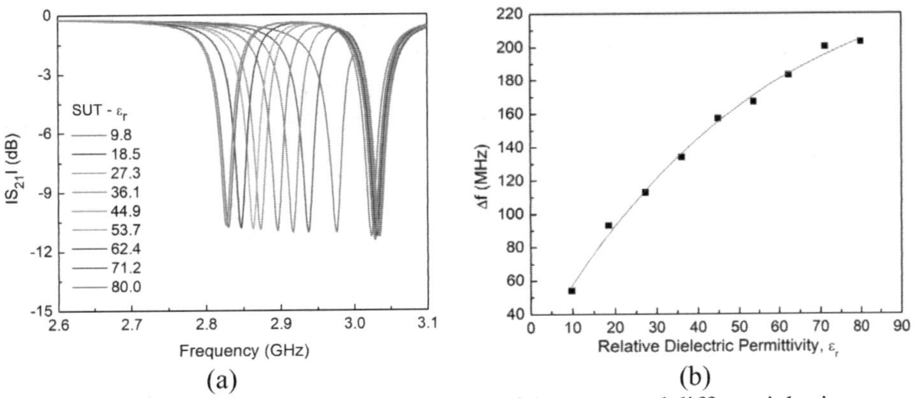

(a) (b)

Figure 4. (a) Simulated frequency responses of the proposed differential microwave sensor of different samples under test values, (b) simulated sensitivity of the differential microwave sensor.

Table 2. Summarized results of the Q-factor analysis.

Permittivity (ε_r)	*Q*-factor
9.8	64.99
18.5	68.24
27.3	68.64
36.1	68.95
44.9	69.18
53.7	69.05
62.4	69.42
71.2	69.86
80.0	70.15

Finally, Table 3 presents a comparison between the performance of the proposed differential microwave sensor and other configurations reported in the literature. In general, the operating principle and the technology of all compared works are the same, *i.e*, the author employs microstrip/planar sensors to measure the dielectric constant of liquid or solid materials in microwave region in all cases. Also, these configurations use the differential method to obtain more stable devices. It is evident that the proposed configuration presents a high sensitivity in comparison with previous reports. Likewise, the proposed configuration could be employed in the characterization of a wide range of materials as it can characterize materials with a relative permittivity of 9.8 to 80.

Table 3. Comparison with other differential planar and microwave sensors. The size is reported in mm, while the sensitivity (S) is reported in MHz/Δεr.

Resonator	Size	S	Range of SUT	Frequency	Reference
Circular SRR	1.62 λg × 1.13 λg	2.89	9.8 – 80	3.0	This work
Split ring resonators	0.57 λg × 0.79 λg	0.79	NA	0.87	[22]
Open complementary SRRs	0.41 λg × 0.29 λg	1.8	28 – 81	0.9	[10]
Dumbbell-shaped defect ground structures stepped	0.87 λg × 0.54 λg	1.03	13 – 81	1.05	[13]
Stepped impedance resonators	NA	0.536	1–11.2	4.0	[12]

*NA No value, * λg is the guided wavelength.

4. Conclusions

A differential microwave sensor based on a pair of transmission lines loaded with split ring resonators has been proposed and numerically investigated in this work. The operating principle of the sensor was analyzed and demonstrated using full-wave simulations. The main relevant characteristics of the proposed structure are that this sensor device has the capability to detect dielectric changes in a wide variety of materials whose relative permittivity is between 9.8 and 80 with high sensitivity. Likewise, the proposed sensor is compact, and the required amount of a given sample is small. Finally, the Quality Factor of this structure was evaluated, and it corroborated that this configuration has a stable resonant peak, which is desirable in real implementations. Therefore, the proposed sensor is an excellent candidate to develop devices for biological, medical, industrial, and other applications where material characteristics are a key aspect.

Acknowledgments

The authors would like to thank the support provided to this work by the Instituto Tecnológico Metropolitano, Colombia, project P21101. The authors F. Umaña-Idarraga, and J. Morales-Guerra acknowledge the support of the Instituto Tecnológico Metropolitano, Colombia, through the Young Researchers program 2021.

References

[1] Chen L F, Ong C K, Neo C P, Varadan V V, Varadan V K 2004 *Microwave Electronics Measurement and Materials Characterization* (Chichester: John Wiley & Sons, Ltd)
[2] Kaatze U 2013 Measuring the dielectric properties of materials. Ninety-year development from low-frequency techniques to broadband spectroscopy and high-frequency imaging *Meas. Sci. Technol.* **24(1)** 012005
[3] Jilnai M, Wen W, Cheong L, Ur Rehman M 2016 A Microwave ring-resonator sensor for non-invasive assessment of meat Aging *Sensors* **16(1)** 52
[4] Raghavan S, Rajeshkumar V 2013 An overview of metamaterials in biomedical applications *Prog. Electromagn. Res. Symp.* PIER (Taipei: PIER) p 368
[5] Acevedo-Osorio G, Reyes-Vera E, Lobato-Morales H 2020 Dual-band microstrip resonant sensor for dielectric measurement of liquid materials *IEEE Sens. J.* **20(22)** 13371
[6] Rodríguez-Moré Z O, *et al.* 2018 Complex dielectric permittivity of rum and its mixtures with methanol, ethanol, and water at frequencies up to 15 GHz *J. Microw. Power Electromagn. Energy* **52(1)** 16

[7] Reyes-Vera E, Acevedo-Osorio G, Arias-Correa M, Senior D E 2019 A submersible printed sensor based on a monopole-coupled split ring resonator for permittivity characterization *Sensors* **19(8)** 1936
[8] Kilpijärvi J, Halonen N, Juuti J, Hannu J 2019 Microfluidic microwave sensor for detecting saline in biological range *Sensors* **19(4)** 819
[9] Su L, Mata-Contreras J, Velez P, Martin F 2016 Splitter/combiner microstrip sections loaded with pairs of complementary split ring resonators (CSRRs): modeling and optimization for differential sensing applications *IEEE Trans. Microw. Theory Tech.* **64(12)** 4362
[10] Velez P, Grenier K, Mata-Contreras J, Dubuc D, Martin F 2018 Highly-sensitive microwave sensors based on open complementary split ring resonators (OCSRRs) for dielectric characterization and solute concentration measurement in liquids *IEEE Access* **6** 48324
[11] Lobato-Morales H, *et al.* 2019 Compact dielectric-permittivity sensors of liquid samples based on substrate-integrated-waveguide with negative-order-resonance *IEEE Sens. J.* **19(19)** 8694
[12] Naqui J, Damm C, Wiens A, Jakoby R, Su L, Mata-Contreras J, Martin F 2016 Transmission lines loaded with pairs of stepped impedance resonators: modeling and application to differential permittivity measurements *IEEE Trans. Microw. Theory Tech.* **64(11)** 3864
[13] Vélez, Muñoz-Enano, *et al.* 2019 Differential microfluidic sensors based on dumbbell-shaped defect ground structures in microstrip technology: analysis, optimization, and applications *Sensors* **19(14)** 3189
[14] Ebrahimi A, Scott J, Ghorbani K 2018 Differential sensors using microstrip lines loaded with two split-ring resonators *IEEE Sens. J.* **18(14)** 5786
[15] Ebrahimi A, Beziuk G, Scott J, Ghorbani K 2020 Microwave differential frequency splitting sensor using magnetic-LC resonators *Sensors* **20(4)** 1066
[16] Munoz-Enano J, Velez P, Barba M G, Mata-Contreras J, Martin F 2020 Differential-mode to common-mode conversion detector based on rat-race hybrid couplers: analysis and application to differential sensors and comparators *IEEE Trans. Microw. Theory Tech.* **68(4)** 1312
[17] Baena J D, *et al.* 2005 Equivalent-circuit models for split-ring resonators and complementary split-ring resonators coupled to planar transmission lines *IEEE Trans. Microw. Theory Tech.* **53(4)** 1451
[18] Smith D R, *et al.* 2000 Composite medium with simultaneously negative permeability and permittivity *Phys. Rev. Lett.* **84** 4184
[19] Castellanos L M, Lopez F, Reyes - Vera E 2016 Metamateriales: principales características y aplicaciones *Rev. la Acad. Colomb. Ciencias Exactas, Físicas y Nat.* **40(156)** 395
[20] Estévez-Hidalgo A, Marante-Rizo F 2019 Antena de microcinta con estructura híbrida metamaterial para 2.4 GHz *Ingeniare. Rev. Chil. Ing.* **27(1)** 22
[21] Catano-Ochoa D, Senior D , Lopez F, Reyes-Vera E 2016 Performance analysis of a microstrip patch antenna loaded with an array of metamaterial resonators *2016 IEEE International Symposium on Antennas and Propagation* IEEE (Fajardo: IEEE) p 281
[22] Velez P, Su L, Grenier K, Mata-Contreras J, Dubuc D, Martin F 2017 Microwave microfluidic sensor based on a microstrip splitter/combiner configuration and split ring resonators (SRRs) for dielectric characterization of liquids *IEEE Sens. J.* **17(20)** 6589

Design of a perfect and multi-resonant metamaterial absorber for electromagnetic energy harvesting applications

F Umaña-Idarraga[1], **D Cataño-Ochoa**[1], **S Montoya-Villada**[1], **C Valencia-Balvin**[1], **and E Reyes-Vera**[1]

[1] Departamento de Electrónica y Telecomunicaciones, Instituto Tecnológico Metropolitano, Medellín, Colombia

E-mail: erickreyes@itm.edu.co

Abstract. In this work, a perfect absorber based on a split ring resonator structure is proposed and numerically analyzed. The software CST STUDIO was employed to carry out the numerical analysis and the optimization of the proposed structure. The electromagnetic properties of the proposed metamaterial cell were analyzed in the first phase of this study demonstrating that such structure resonates at 2.4 GHz and 4.2 GHz simultaneously. In fact, this structure has negative permittivity and permeability in these two bands. The optimization process has led us to obtain a compact resonator, which has a total size of 15 mm × 15 mm. Subsequently, the capacity of this structure as an absorber of electromagnetic energy is analyzed. The obtained results reveal that this structure has absorption efficiencies of 98.2% and 99.7% for the first and second bands respectively. Also, other characteristic parameters were evaluated. This shows that the proposed structure has a high electrical performance and can be used for the collection of electromagnetic energy, which can be used to power wireless sensor networks.

1. Introduction

Recently there has been an increase in the number of wireless sensing networks (WSN), which has been mainly due to the emergence of new technologies such as the internet of things (IoT), 5G networks, and the interest in automating different industrial processes [1–3]. Due to the wireless and electronic sensors must be installed in the outdoors in most cases, it has become necessary to implement energy harvesting systems to obtain energy from the environment and supply power to low-energy electronic devices or recharge their batteries. Different energy harvesting mechanisms such as thermal energy, wind energy, solar energy, and electromagnetic energy (EM) have been explored [4]. This last alternative has attracted a lot of attention in recent years because EM energy is omnipresent today in our environment due to the abundance of EM signals available from different sources such as mobile stations, Wi-Fi signals, or television antennas [4].

In addition, they open the possibility of supplying power wirelessly, *i.e*, build wiring-free power distribution systems. For this reason, the design of electromagnetic energy collectors plays a fundamental role in the capture and conversion of this energy to be used in sensing nodes [5,6]. To solve this problem, several alternatives have been explored. The most used mechanism are antenna arrays [7], rectennas [8], and metamaterial absorbers [6,9]. The metamaterial absorbers are composed of unit cells arranged periodically in a 2-D plane. These unit cells have unique characteristics such as negative permittivity and negative permeability [10–12]. In fact, the unique characteristics of these metamaterial cells have aroused strong interest in other applications such as antenna miniaturization [13,14] sensing devices [15–17], and among others [12,18].

The main advantage of a metamaterial absorber energy harvesting system is that this kind of structure is smaller in comparison with conventional systems based on antennas. Likewise, these structures exhibit a strong resonance and a purely effective surface resistance, which facilities electromagnetic wave absorption [4]. On the other hand, the metamaterial absorber receives the EM wave from the environment at the resonant frequencies and stored the energy (electric and magnetic) due that these structures have capacitance and inductance associated [4,19]. Thus, the maximal absorption is achieved when the structure impedance matches that of free space. After that, the stored energy is dissipated through the dielectric substrate or a load. Then, the performance of the absorber can be evaluated based on the absorption efficiency, the harvesting efficiency, and its conversion efficiency.

Therefore, we propose to design a double split-ring resonator with a compact size to operate at 2.4 GHz and 4.2 GHz simultaneously in this work. In comparison with other previous studies, the proposed structure has some important advantages. Firstly, the unit cell resonates at two different frequencies. Secondly, the proposed structure has the capability to absorb the EM radiation with higher efficiency, this has an efficiency of more than 98% in both bands.

2. Methodology

Figure 1 shows the proposed metamaterial unit cell; this metamaterial cell consists of three concentric rings. The bigger resonator has a square shape with one gap in the middle of each side of two adjacent lines, while the other ones correspond to conventional split ring resonators (SRRs) with a circular shape, which are rotated -45 degrees to induce a high asymmetry in the structure. The ring resonators are made of copper lines with a metal thickness of 35 μm, and the bottom layer is a continuous layer of copper. On the other hand, the three-ring SRR structure was designed and simulated on a RO4003C substrate (Rogers Corp.), which has a relative permittivity of 3.55, a thickness of 1.524 mm, and a loss tangent of 0.0021. The total size of this unit cell is 15 mm × 15mm and operates at 2.4 GHz and 4.2 GHz simultaneously; Table 1 summarized the most important geometrical parameters of the unit cell.

Table 1. Metamaterial unit cell design parameters.

Parameter	Value (mm)
L	14.00
Wc	1.00
Wr	1.00
W	1.00
g	0.70
g1	0.65
K	15.0
α	−45°

The performance of the proposed unit cell was numerically evaluated using CST Microwave Studio, which is a software based on the standard finite-difference time-domain (FDTD) method. To carry out the simulation was important to choose the correct boundary conditions. Then, the unit cell was placed between two floquet ports to study its electromagnetic response, while perfect electric and perfect magnetic boundary conditions were imposed on the metamaterial unit cell (see Figure 1(b)); from Figure 1(b), the structure was excited using a uniform plane wave, which is propagated on the z-axis with open end boundary condition. Likewise, a perfect electrically conducting (PEC) was used in z-direction, while the unit cell condition was applied in x-direction and y-direction. Therefore, the electric field (E-field) of the incident wave is polarized along x-axis, while the magnetic field (H-field) is polarized along y-axis.

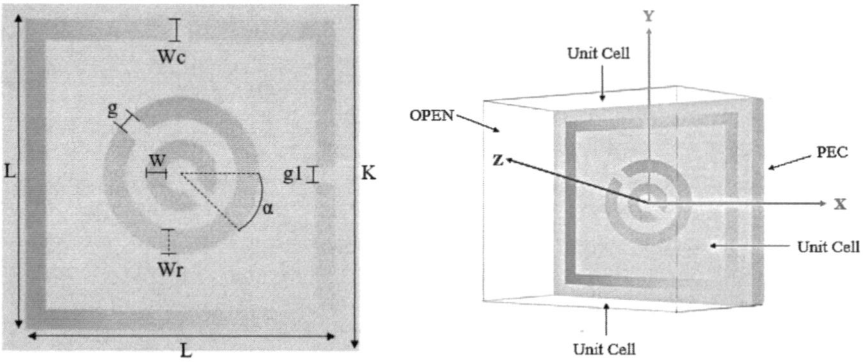

(a) (b)

Figure 1. (a) Schematic of the proposed metamaterial unit cell, (b) boundary condition defined in CST Microwave STUDIO for the studied metamaterial unit cell.

The S_{11} and S_{21} parameters of the proposed unit cell were analyzed, and these results were used to determine the electromagnetic behavior between 2 GHz and 5 GHz; then, the Nicolson Rose Weir (NRW) method was employed to extract and get the permeability and permittivity of the proposed structure from the S-parameters [11,20]. These parameters can be obtained easily from the following Equation (1), and Equation (2) [20].

$$\mu_r = \frac{2c(1-S_{21}+S_{11})}{j\omega d(1+S_{21}-S_{11})}, \tag{1}$$

$$\varepsilon_r = \mu_r + j\frac{2cS_{11}}{\omega d}, \tag{2}$$

where c is the speed of light, μ_r is the relative magnetic permeability, ε_r is the relative dielectric permittivity, ω is the angular frequency, S_{11} represents the reflected power, S_{21} represents the transmitted power, and d is the slab thickness. A similar procedure was performed to analyze a 3×3 array configuration. The structure of the proposed array is illustrated in Figure 2; here, the unit cells are separated from each other by a distance $\Lambda = 15$ mm.

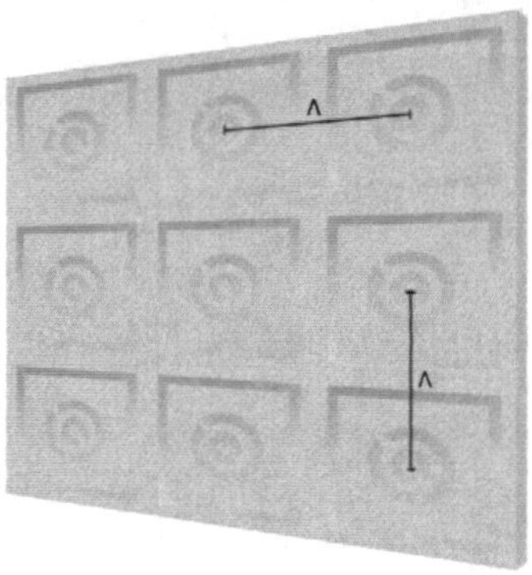

Figure 2. Schematic of the proposed 3×3 array configuration of unit cells.

3. Results and discussion

The metamaterial unit cell was studied using the methodology described in the above section. Figure 3 shows the obtained S-parameters (the reflection coefficient, S_{11}, and the transmission coefficient, S_{21}) from 2 GHz to 5 GHz. The results reveal that the proposed structure has a strong resonance at two different bands. The first one is 2.4 GHz, where the S_{11} parameter reaches a magnitude of -17.6 dB, while the magnitude of the S_{21} parameter is -1.07 dB. Similarly, this unit cell resonates at 4.2 GHz.

In fact, the S_{11} and S_{21} parameters at this frequency present a magnitude equal to –29.8 dB and -0.29 dB respectively. Another important characteristic of this kind of electromagnetic absorber is the bandwidth because it determines the range of frequencies in which the proposed structure can work. Then, the bandwidth of the proposed absorber was calculated at both resonances. Thus, the bandwidth in the first band was 36 MHz, while the second band reported higher bandwidth. In fact, this band has a bandwidth equal to 125.65 MHz Thus, the second band has at least three times more bandwidth than the first band.

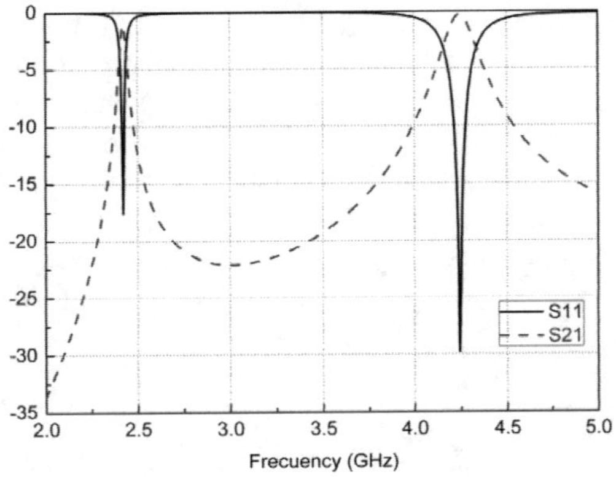

Figure 3. Magnitude of the S_{11} and S_{21} parameters of the proposed metamaterial unit cell. The continuous line indicates the magnitude of parameter S_{11}, and the dashes indicate the magnitude of parameter S_{21}.

To confirm that the proposed unit cell has left-handed frequency bands, the effective material electromagnetic parameters were obtained using the Equation (1) and Equation (2). Figure 4(a) and Figure 4(b) shows the obtained retrieved effective real and imaginary part of the permittivity and permeability as a function of the operating frequency. The numerical results corroborate that the proposed metamaterial unit cell exhibit negative permeability and permittivity at 2.4 GHz and 4.2 GHz simultaneously. So, this structure shows double negative property and can be used as double negative (DNG) medium [21].

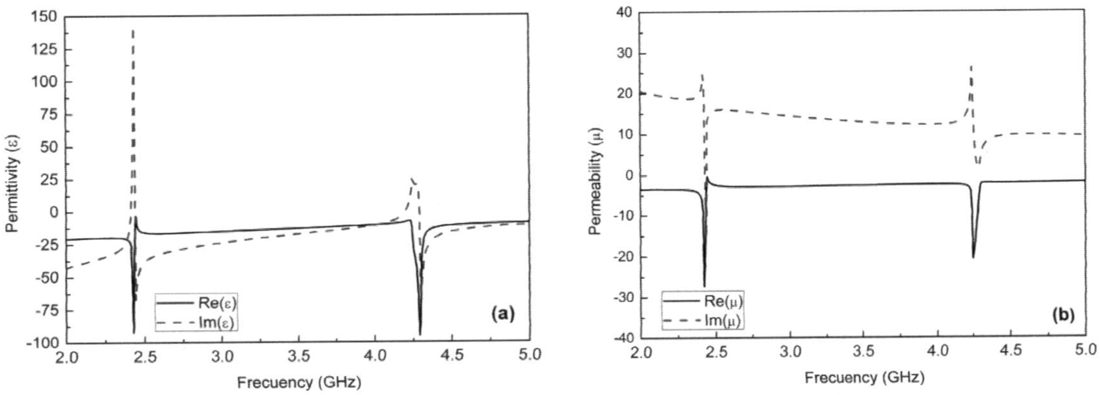

(a) (b)

Figure 4. Simulation results of the (a) permittivity (b) permeability. The obtained results were obtained using the NRW method. Solid curves represent the real part, while the dashed lines show the behavior of the imaginary part.

Finally, the capability to absorb the electromagnetic energy of the proposed structure was evaluated. Figure 5 illustrates the numerical values of the absorbance A(w), the transmittance T(w), and the reflectance R(w). Here, the absorbance can be calculated using the Equation (3) [6].

$$A(\omega) = 1 - R(\omega) - T(\omega), \tag{3}$$

where the frequency-dependent reflectance and transmittance are defined as $R(\omega) = |S_{11}|^2$ and $T(\omega) = |S_{21}|^2$ respectively [6]. Figure 5 reveals that the absorbance of the proposed metamaterial absorber has an absorbance of 98.2% and 99.7% at 2.4 GHz and 4.2 GHz respectively. Likewise, the reflectance is 0.13% and 0.032% in the first and second operating bands of the proposed absorber. Finally, the transmittance is near zero across the analyzed frequency range. The proposed absorber can be employed to collect electromagnetic energy in one of the most used electromagnetic bands because most current wireless networks radiate close to 2.4 GHz. Therefore, this absorber can surely be used in the capture of high energy rates, which at the same time can be used to power wireless networks, batteries, or other electronic components [4-7].

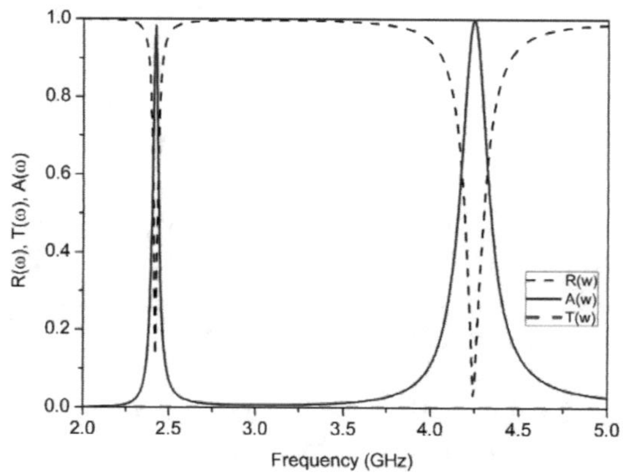

Figure 5. Numerical results of the reflectance, transmittance, and absorbance for the proposed metamaterial absorber.

4. Conclusions

In summary, we proposed a novel metamaterial unit cell based on three concentric split-ring resonators, which can be used to collect electromagnetic energy. First, the capability of this structure to absorb the incident electromagnetic energy at 2.4 GHz and 4.2 GHz was numerically demonstrated. The numerical results demonstrate that the proposed metamaterial structure presents a high absorbance in both bands. Likewise, this metamaterial absorber offers several advantages over conventional electromagnetic energy collection systems because it is compact, thin, and offers a high absorption coefficient in multiple bands simultaneously. Thus, the proposed structure is an ideal candidate for energy harvesting applications due that it offers high performance to collect and store electromagnetic energy.

Acknowledgments

The authors would like to thank the support provided to this work by the Instituto Tecnológico Metropolitano, Colombia, project P21101. The author Umaña-Idarraga F acknowledges the support of the Instituto Tecnológico Metropolitano, Colombia, through the Young Researchers program 2021, and S. Montoya-Villada acknowledges the support of Ministerio de Ciencia Tecnología e Innovación, Colombia, through the program "Convocatoria fortalecimiento de vocaciones y formación en CTeI para la reactivación económica en el marco de la postpandemia 2020".

References

[1] Reyes-Vera E, Senior D, Luna-Rivera J M, López-Giraldo F 2018 Advances in electromagnetic applications and communications *TecnoLógicas* **21(43)** 9

[2] Botero Valencia J, Castaño Londoño L, Marquez Viloria D 2019 Trends in the internet of things *TecnoLógicas* **22(44)** 1

[3] Liu D, Hong W, Rappaport T S, Luxey C, Hong W 2017 What will 5G Antennas and propagation be *IEEE Trans. Antennas Propag.* **65(12)** 6205

[4] Amer A A G, Sapuan S Z, Nasimuddin N, Alphones A, Zinal N B 2020 A Comprehensive review of metasurface structures suitable for RF energy harvesting *IEEE Access* **8** 76433

[5] Zhang Y, Zhao J, Cao J, Mao B 2018 Microwave metamaterial absorber for non-destructive sensing applications of grain *Sensors* **18** 1912

[6] Cheng Y Z, Cong Fang, Zhe Zhang, Bin Wang, Junfeng Chen, Gong R Z 2016 A compact and polarization-insensitive perfect metamaterial absorber for electromagnetic energy harvesting application *Progress in Electromagnetic Research Symposium IEEE* (Shanghai: IEEE) p 1910

[7] Mavaddat A, Armaki S H M, Erfanian A R 2015 Millimeter-wave ener hagyrvesting using 4×4 microstrip patch antenna *Array IEEE Antennas Wirel. Propag. Lett.* **14** 515

[8] Wagih M, Weddell A S, Beeby S 2020 Rectennas for radio-frequency energy harvesting and wireless power transfer: a review of antenna design *IEEE Antennas Propag*. Mag. **62** 95

[9] Chomtong P, Rakdanklang K, Krachodnok P, Bandudej K, Akkaraekthalin P 2021 A Dual-band metasurface reflector using ring resonator with interdigital capacitor *IEEE Access* **9** 62074

[10] Smith D R, Pendry J B, Wiltshire M C K 2004 Metamaterials and negative refractive *Index Science* **305(5685)** 788

[11] Smith D R, Vier D C, Koschny T, Soukoulis C M 2005 Electromagnetic parameter retrieval from inhomogeneous metamaterials *Phys. Rev. E* **71(3)** 036617

[12] Castellanos L M, Lopez F and Reyes-Vera E 2016 Metamateriales: principales características y aplicaciones *Revista de la Academia Colombiana de Ciencias Exactas, Físicas y Naturales* **40(156)** 395

[13] Catano-Ochoa D, Senior D E, Lopez F, Reyes-Vera E 2016 Performance analysis of a microstrip patch antenna loaded with an array of metamaterial resonators *IEEE International Symposium on Antennas and Propagation* (Fajardo: IEEE) p 281

[14] Rao S M, Basarkod P 2020 A novel complementary slotted split ring resonator loaded truncated arc patch antenna with enhanced performance *Prog. Electromagn. Res. C* **101** 203

[15] Acevedo-Osorio G, Reyes-Vera E, Lobato-Morales H 2020 Dual-band microstrip resonant sensor for dielectric measurement of liquid materials *IEEE Sens. J.* **20(22)** 13371

[16] Reyes-Vera E, Acevedo-Osorio G, Arias-Correa M, Senior D 2019 A submersible printed sensor based on a monopole-coupled split ring resonator for permittivity characterization *Sensors* **19(8)** 1936

[17] Raghavan S, Rajeshkumar V 2013 An overview of metamaterials in biomedical applications *Prog. Electromagn. Res. Symp.* (Taipei: PIERS) p 368

[18] Gil M, Bonache J, Martín F 2008 Metamaterial filters: a review *Metamaterials* **2(4)** 186

[19] Baena J D, Bonache J, Martin F, Sillero R, Falcone F, Lopetegi T Laso M, Garcia-Garcia J, Gil I, Portillo M F, Sorolla M 2005 Equivalent-circuit models for split-ring resonators and complementary split-ring resonators coupled to planar transmission lines *IEEE Trans. Microw. Theory Tech.* **53(4)** 1451

[20] Numan A B, Sharawi M S 2013 Extraction of material parameters for metamaterials using a full-wave simulator *IEEE Antennas Propag. Mag.* **55(5)** 202

[21] Smith D R, Smith D R, Padilla W J, Padilla W J, Vier D C, Vier D C, Nemat-Nasser S C, Nemat-Nasser S C, Schultz S, Schultz S 2000 Composite medium with simultaneously negative permeability and permittivity *Phys. Rev. Lett.* **84(18)** 4184

Fluid-dynamic study of the behavior of the air inside a textile Stenter

J W Parra[1], M B Quadri[2], and D C Rodríguez[1]
[1] Grupo de Investigación en Química Básica Aplicada, Universidad Francisco de Paula Santander, San José de Cúcuta, Colombia
[2] Laboratório de Sistemas Porosos, Universidade Federal de Santa Catarina, Florianópolis, Brasil

E-mail: johnwilmerpl@ufps.edu.co

Abstract. In the textile industry, drying is one of the most important processes. This process requires large investments and high energy consumption, which generates high costs for companies in this sector. In this work, a modeling of the behavior of the air was carried out in a textile Stenter, under real operating conditions through the development of fluid-dynamic simulations. For the computational modeling of the problem, a 3D geometry was constructed based on measurements taken from an injector of a textile Stenter. The standard k-ε turbulence model was used in the turbulent flow solution. The equations of the model were solved numerically using the finite element method. The standard k-ε turbulence model proved to be a model capable of reproducing the behavior of the air in the injectors of the textile Stenter.

1. Introduction

In the textile industry, drying is one of the most important processes. From a physical point of view, the drying of the fabric can be defined as a balance between the transfer of heat from the air stream to the surface of the fabric and the transfer of moisture from the surface of the fabric to the air stream. This process requires large investments and high energy consumption, which generates significant costs for companies in this sector. In industrialized countries, around 7% to 15% of the energy generated for industrial purposes is generally used for drying with relatively low thermal efficiency [1].

Therefore, it is of great interest to find ways to reduce energy demand without damaging the final characteristics of the fabric. Textile Stenters are thermal equipment used to dry fabrics in a continuous process. In the drying chambers, hot air flows over the fabric and dries it by convection with the exchange of heat and mass. The exhaust system removes the vapors resulting from the drying of the fabric, reducing the humidity of the air in the system and, consequently, increasing the drying efficiency in the process [2].

In the literature there are several studies on turbulent air jets, in which they carry out numerical studies, comparing different types of turbulence models with experimental values, this in order to study the heat and mass transfer applied in industrial processes, including drying. For example, Shi Y, *et al*, [3] studied the used of the standard turbulence model and the Reynolds stress model (RSM), model to calculate the heat transfer for a single semi-confined turbulent jet, which impinges on a flat plate, using different airflow parameters and different geometric parameters. For low heat flux values (1000 W/m^2 - 6000 W/m^2), the physical properties of air were independent of temperature and heat flux; however, for high heat flux (50000 W/m^2), they observed that the physical properties of air were dependent on temperature.

Colucci and Viskanta [4] studied the effect of the geometry of a hyperbolic nozzle on local heat transfer coefficients, for this, they used a thermochromic liquid crystal technique to visualize and record isotherms on a uniformly heated impact surface. The authors concluded that the heat transfer coefficients at the site of incidence of the confined jets are more sensitive to the Reynolds number and the space between the nozzle and the plate, than the respective coefficients for the unconfined jets. Other authors who used the liquid crystal technique were Lee J. and Lee S [5]. They studied the effects of the nozzle outlet configuration on improvements in heat transfer, experimenting with three different types of nozzles.

Wang and Mujumdar [6] used 5 low Reynolds number models to study a two-dimensional turbulent jet, most of the models showed good agreement with the experimental data, both in the stagnant regions and in the jet walls. Gulati P, *et al* [7]. Studied the effects of the nozzle shape, the space between the plate and the jet, and Reynolds number, on the local heat transfer distribution of an air jet on a smooth and flat surface, for this they used three nozzles with different geometries: circular, square and rectangular, each with an equivalent diameter of 20 mm.

In this work the objective was to study the behavior of the air inside the drying chambers of a textile Stenter under real operating conditions, by obtaining speed fields, turbulent kinetic energy profiles and dissipation rate of turbulent air in the injectors, with in order to relate them to the phenomena of heat and mass transfer.

2. Materials and methods

This section presents the mathematical methodologies used to carry out the study this work.

2.1. Mathematical formulation

The simulations were carried out in COMSOL Multiphysics® version 4.4, a phenomenological simulation and modeling software. COMSOL Multiphysics® internally compiles the set of equations that represent the model in its entirety, generating a system of linear equations. The resolution of the model is obtained by the finite element method (FEM), which is numerical procedure used to solve systems of partial differential equations (PDE), and consists of dividing the domain of the solution into regions of simple form or by elements. Then, a rough solution for electronic data processing can be developed for each of these elements. The total solution is generated by joining the individual solutions and, at the same time, continuity at the boundaries between the elements is ensured [8]. For this study, the standard *k-ε* turbulence model was used, which is the most used for computational simulations, since it presents good convergence and requires relatively few computational resources.

2.2. k-ε turbulence model

Turbulence is a phenomenon in which fluid particles mix in a non-linear way, which is described through various models of turbulence; the turbulence model (k-ε), used in this work is governed by two parameters: k related to the kinetic energy of the turbulence (defined as the variation of the fluctuations in speed) and ε the dissipation of the turbulence eddy (the rate at which speed fluctuations dissipate) [9]. Turbulent stresses are modeled through turbulent viscosity that include the parameters k and ε. For the steady state, the values of k and ε come directly from the differential equations for turbulence kinetic energy transport and the turbulence dissipation rate (see Equation (1) and Equation (2), respectively [10]).

$$\rho u \cdot \nabla k = \nabla \cdot \left(\left(\mu + \frac{\mu_T}{\sigma_k} \right) \nabla k \right) + P_k - \rho \varepsilon, \tag{1}$$

$$\rho u \cdot \nabla\varepsilon = \nabla \cdot \left(\left(\mu + \frac{\mu_T}{\sigma_\varepsilon} \right) \nabla\varepsilon \right) + C_{\varepsilon 1} \frac{\varepsilon}{k} P_k - C_{\varepsilon 2} \rho \frac{\varepsilon^2}{k}, \quad (2)$$

where u is the velocity, ρ the density and μ the viscosity of the fluid; $C_{\varepsilon 1}$, $C_{\varepsilon 2}$ and C_μ are constants equal to 1.44, 1.92 and 0.09 respectively. $\sigma_k = 1.0$ is the turbulent Prandlt number for k and σ_ε is theturbulent Prandlt number for ε. The Equation (3) calculate P_k is the turbulent kinetic energy production due to buoyancy and viscous forces [10]).

$$P_k = \mu_T \left(\nabla u \colon \left(\nabla u + (\nabla u)^T \right) - \frac{2}{3} (\nabla \cdot u)^2 \right) - \frac{2}{3} \rho k \nabla \cdot u, \quad (3)$$

The Equation (4) calculate Turbulent viscosity μ_T is modeled by [10].

$$\mu_T = \rho C_\mu \frac{k^2}{\varepsilon}. \quad (4)$$

2.3. Simulation parameters

During the development of this work, a technical visit was made to a company in the textile sector that uses direct heating textile Stenters; on this visit, we were able to access the textile Stenter injectors (see Figure 1) and their measurements were also taken with a tape measure.

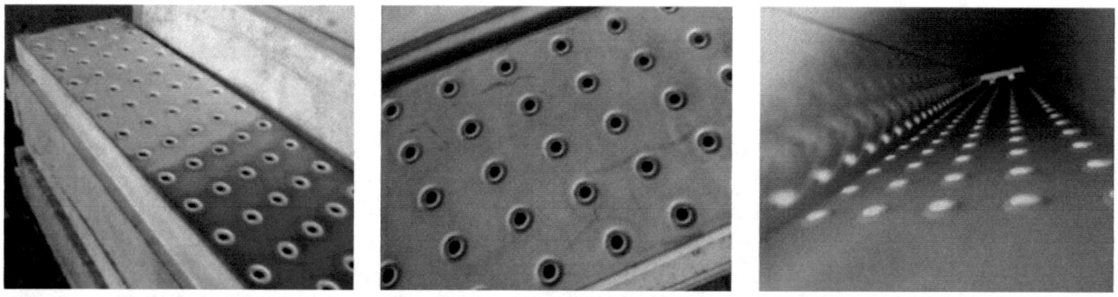

Figure 1. Photographs of the injector of a textile Stenter.

With the dimensions measured to the injectors during the technical visit, it was possible to build a geometry closer to the real one; thus, a 3D geometry was constructed consisting of two injectors (upper and lower) and a rectangular section volume in the middle of them, which represents the fabric in motion at a translation speed of the fabric (V_{fabric}) of 0.4 m/s entering longitudinally in the dryer (x-axis), as shown in Figure 2 and at a temperature of 30 °C. The following physical properties were considered for the cotton fabric: density of 1500 kg/m^3, specific heat of 1339.78 J/kg K and thermal conductivity of 0.029 W/(m K) values used by [11]. In this case, the air enters the injectors at a speed (V_{air}) of 5.0 m/s and 180 °C of temperature, operating parameters used in industry [2,11]. The air injector is made up of a duct 2.82 m long, 0.31 m wide and 0.06 m high at one end and 0.19 m at the other (see Figure 2).

Each injector has 170 nozzles (distributed in 5 rows of 34, not paired laterally). The nozzles used in textile Stenters have a hemi-toroidal shell rim; the hemi-toroidal shell shape of the nozzle aims to distribute the jet of air more efficiently over the tissue surface. The nozzles have an inlet diameter (D_{nozzle}) of 0.015 m and a height (A_{nozzle}) of 0.005 m. The space between the injectors and the surface of the cotton fabric is (H) is 0.07 m. To represent the cotton fabric, a flat plate of 2.65 m long, 0.31 m wide and 0.0015 m thick was constructed.

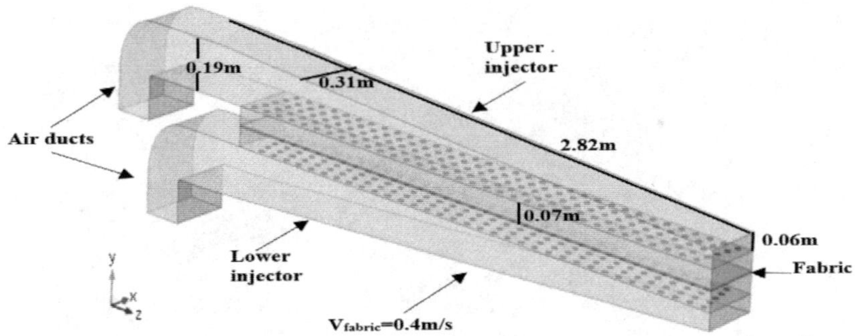

Figure 2. 3D geometry of the upper and lower injectors of a direct heating textile stenter.

2.4. Geometry and discretization mesh

The construction of the mesh is one of the most important aspects to develop computational models, since the good construction of the mesh depends to a great extent on the convergence of a simulation, as well as the convergence time; due to the complexity of the domain, since it is composed of several subdomains, a mesh of eight hexahedral layers and 1018054 elements was built, which can be seen in Figure 3.

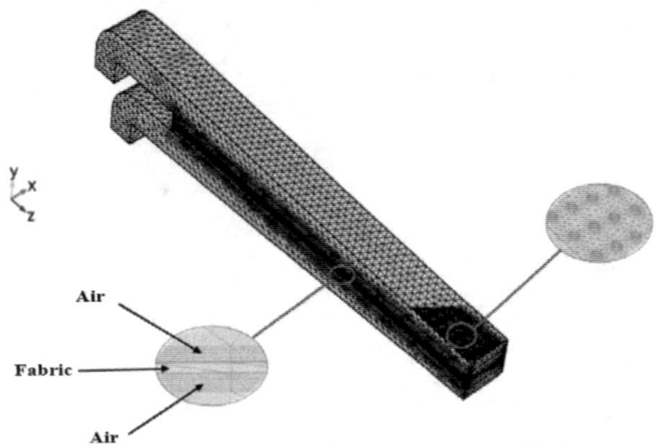

Figure 3. Generated mesh of a pair of injectors.

3. Results and discussions

The following is the analysis of the results obtained during the dynamic fluid study of the behavior of the air inside the textile Stenter used in the textile industry.

The objective to qualify and quantify the air flow in the region of the injectors of the dryer under study, by analyzing the velocity profiles, the streamlines and the turbulence parameters obtained from the k-ε turbulence model. As mentioned in section 2.3, velocities of 5.0 m/s were adopted for the air inlet velocity in the reference injectors and 0.4 m/s for the fabric displacement velocity.

In Figure 4, which shows the flow of air developed at steady state, it is possible to discern the velocity fields along a longitudinal section plane of the injectors, including the tissue surface. When analyzing the behavior of the air in the injector, it is observed that the highest velocity values are reached at the end nozzles where values between 8 m/s and 10 m/s are observed, showing that, in these areas of the injectors, the air reaches up to twice the inlet velocity. In the nozzles of the central region, the velocity reaches values between 4 m/s and 7 m/s.

It can also be observed that, together with the extreme nozzles up to the surface of the tissue, the jets promote an intensification of the air velocity, inducing the appearance of vortices that characterize the behavior of the turbulent flow.

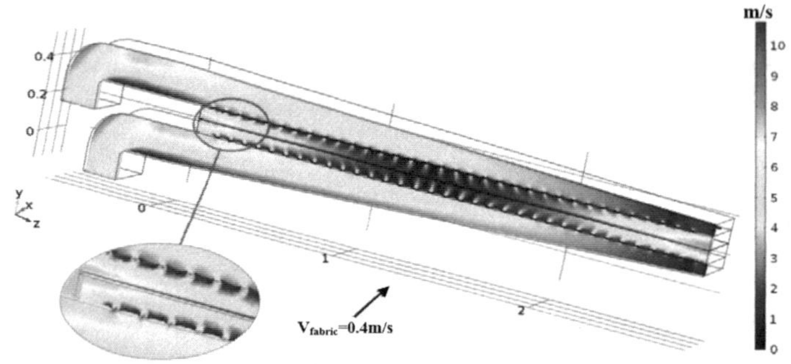

Figure 4. Velocity field in the injectors.

Figure 5, Figure 6 and Figure 7 present the flow velocity fields and streamlines in 2D graphs of longitudinal (z-y) and transverse (x-y) sections. Figure 5(a) corresponds to the injector cut in the (z-y) plane for $x = 0.155$ and Figure 5(b) corresponds to the injector cut in the (x-y) plane for $z = 0.320, z = 1.445$ and $z = 2.645$.

The air recirculation areas near the inlet of the ducts that lead the air to the injectors are shown in Figure 6. Note that the velocity values in these regions are high; according to [12], this is due to the fact that recirculation restricts the passage of air freely, forming a stagnant region. Thus, due to the laws of conservation of mass and motion, the air accelerates in this region and moves towards the interior of the injector, subsequently losing speed.

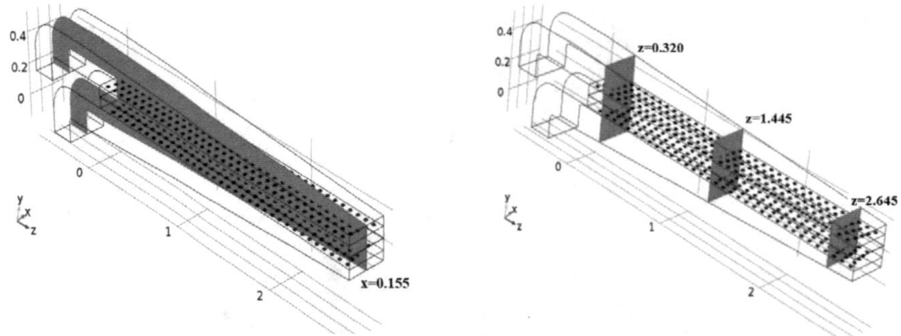

(a) (b)

Figure 5. Injector cuts (a) in the (z-y) plane and (b) in the (x-y) plane.

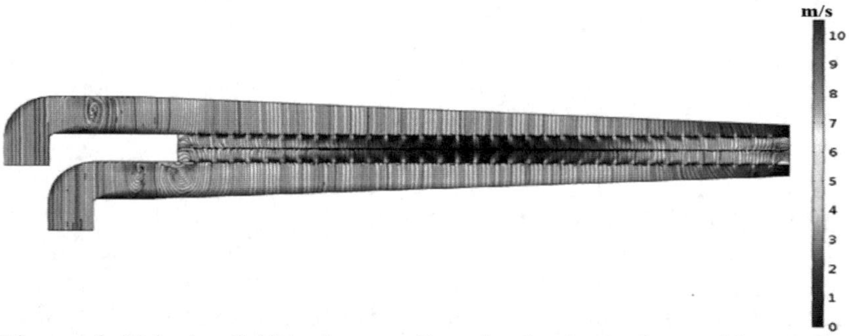

Figure 6. Velocity field and streamlines in the (z-y) plane with cutoff at $x = 0.155$.

In Figure 7 it can be seen that the current lines inside the injectors are quite orderly and uniform, as they leave them and meet the tissue, which is in motion, the lines begin to behave in a disorderly way and form vortices displaced in the direction of tissue movement.

Journal of Physics: Conference Series **2118** (2021) 012006 doi:10.1088/1742-6596/2118/1/012006

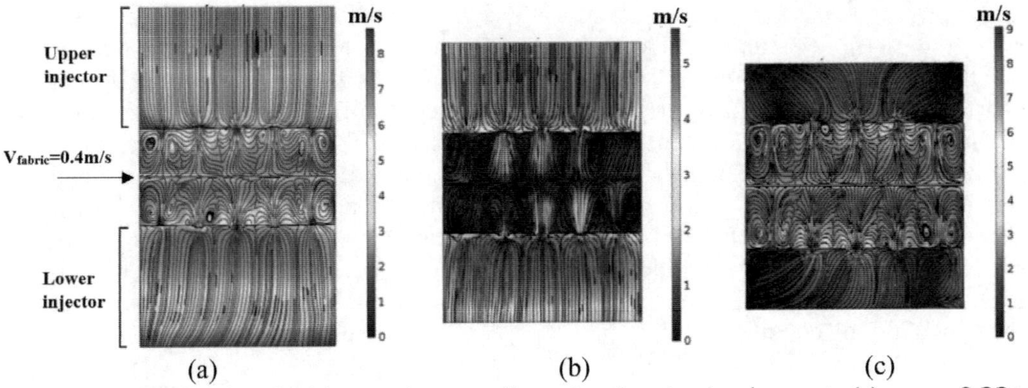

Figure 7. Field of velocities and streamlines in the (x-y) plane at (a) z = 0.32, (b) z = 1.445, and (c) z = 2.645.

In Figure 8, it can be seen that the turbulent kinetic energy has the highest values at the ends of the injectors; this is because the greatest flow accelerations occur in these regions. In the central region of the injectors, the turbulent kinetic energy tends to be lower.

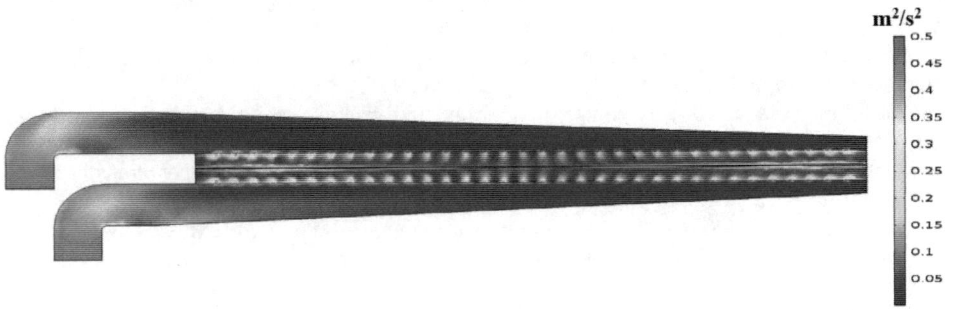

Figure 8. Turbulent kinetic energy in the (z-y) plane at x = 0.155.

The rate at which energy dissipates is shown in Figure 9. It is observed that the behavior is very similar to that of turbulent kinetic energy, that is, the highest values are found at the ends of the injectors and are lowest in the central region.

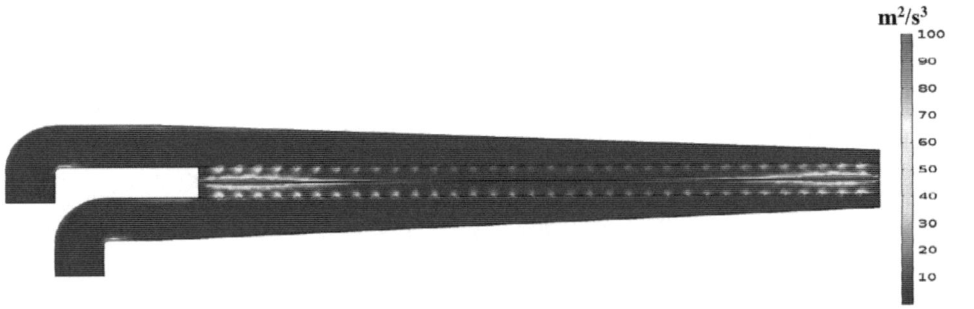

Figure 9. Turbulence dissipation rate in the (z-y) plane at x = 0.155.

A similar behavior was observed by Janzen J G, *et al* [13]. Who studied the turbulence properties of a viscous fluid through the oscillation of two grids; they observed the existence of two regions: a region closes to the grids, where turbulence-producing effects are relevant (eventually also the advection resulting from the jets that form on the grids); and a second region (further from the gratings) in which there is a decrease in turbulent kinetic energy.

4. Conclusions

Regarding the air flow patterns inside the textile Stenter, the k - ε turbulence model allowed to consistently simulate the fluid-dynamic behavior of the air according to the operating conditions. By studying the behavior of the air in the injectors, the velocity profiles could be obtained, observing a non-uniform distribution of the intensity of the jets in the nozzles along them. The speed reached maximum values at the ends of the injectors. The formation of eddies and stagnant regions with behavior physically consistent with the fluid-dynamic principles applied to the geometry under study, was also evidenced. It is concluded, therefore, that the distribution of the air inside the dryer is not uniform for the geometry studied, which does not allow to support the statements, generally made by manufacturers of these equipment with geometries similar to the one studied, regarding the homogeneity of drying conditions.

References

[1] Oktay Z, Hepbasli A 2002 Performance evaluation of a heat pump assisted mechanical opener dryer *Energy Conversion and Management* **44** 1193
[2] Ferraz A D 2010 *Rio Oil & Gas Expo and Conference 2010* (Rio de Janeiro: Instituto Brasileiro de Petróleo, Gás e Biocombustíveis)
[3] Shi Y, Ray M B, Mujumdar A S 2002 Computational study of impingement heat transfer under a turbulent slot jet *Ind. Eng. Chem. Res.* **41** 4643
[4] Colucci D W, Viskanta R 1996 Effect of nozzle geometry on local convective heat transfer to a confind impinging air jet *Experimental Thermal and Fluid Science* **13** 71
[5] Lee J, Lee S J 2000 The efect of nozzle configuration on stagnation región heat transfer enhancement of axisymmetric jet impingement *Int. J. Heat Mass Trans.* **43** 3497
[6] Wang S J, Mujumdar A S 2005 A comparative study of five low Reynolds number k– ε models for impingement heat transfer *Applied Thermal Engineering* **25** 31
[7] Gulati P, Katti V, Prabhu S V 2008 Influence of the shape of the nozzle on local heat transfer distribution between smooth flat surface and impinging air jet *Int. J. of Ther. Scie.* **48** 602
[8] Chapra S C, Canale R P 2008 *Métodos Numéricos para Ingenieros, 5ª edición* (Mexico: McGraw Hill)
[9] Heuert J, Khatchatourian O 2007 *Congresso Nacional de Matemática Aplicada e Computacional* (Florianópolis: Sociedade Brasileira de Matemática Aplicada e Computacional)
[10] Jones W P, Launder B E 1972 The prediction of laminarization with a two-equation model of turbulence *International Journal of Heat and Mass Transfer* **15** 301
[11] Santos R M, Parra J W, Quadri M B, Rocha I C 2015 Study of drying and consumption of natural gas in a textile stenter of direct heating *Dry Technol.* **33** 37
[12] Galarça M M 2004 *Análise Numérica para Modelos de Turbulência k-ω E SST/ k-ω para O Escoamento de Ar No Interior de Uma Lareira de Pequeno Porte. Modelagem Da Turbulência* (Porto Alegre: Universidade Federal do Rio Grande do Sul)
[13] Arantes E J, Porto R M, Gulliver J S, Lima A, Schulz H E 2010 Lower nappe aeration in smooth channels: experimental data and numerical simulation *Anais da Academia Brasileira de Ciências* **82** 521

Journal of Physics: Conference Series **2118** (2021) 012007 doi:10.1088/1742-6596/2118/1/012007

Analysis of heat fluxes in ceramic block type building pieces

M S Narváez-Ortega[1,2], J Sánchez-Molina[1], and J V Sánchez-Zúñiga[1]
[1] Grupo de Investigación en Tecnología Cerámica, Universidad Francisco de Paula Santander, San José de Cúcuta, Colombia
[2] Grupo de Investigación en Arquitectura y Materiales Alternativos, Universidad Francisco de Paula Santander, San José de Cúcuta, Colombia

E-mail: stephannynarvaezortega@outlook.com

Abstract. In Norte de Santander, Colombia, there is a ceramic cluster made up of 59 companies dedicated to the production of construction materials derived from clay, a vernacular material with a wide extraction potential in the region. Fired clay products in different hollow block presentations represent 15% of industrial production, these construction pieces are low cost and show a high demand in the local masonry market, therefore, their thermo-physical characteristics largely build the thermal envelope of "Cucuteña". This research comparatively evaluates the thermal behavior of four types of ceramic blocks with different perforations, applying the finite element method in thermal simulations that consider a conductivity of 0.407 W/m°C in the ceramic to establish the influence of the shape on the temperature distribution and heat flow profiles of each piece subjected to environmental conditions of a warm semi-arid climate, with average maximum temperatures of 33 °C and an average maximum solar radiation of 796.8 Wh/m^2 under extreme climatic conditions in the city of San Jose de Cúcuta, Colombia. The results of the study allow identifying the ceramic block of 6 rectangular holes as the product that, due to its physical characteristics, presents a better alternative to be applied in constructions that seek thermal efficiency.

1. Introduction

In the department of Norte de Santander, the richness of the location defines the constructive materiality with which its architecture is built, the clay, as a material wealth, is a vernacular resource with excellent physical and chemical characteristics for the manufacture of ceramic pieces for construction. Therefore, in this northeast area of Colombia there is an industrial cluster that is responsible for its exploitation and transformation, bringing together 59 companies dedicated to the production of construction ceramics [1], which make up one of the most important subsectors of the local economy, representing 21% of the manufacturing sector, producing 38% of employment and contributing 12.8% of the GDP in the region [2,3], the goods derived from this production process generate a large supply of materials for the construction sector such as blocks, bricks, tiles, slabs, among others.

The present research seeks to evaluate the thermal behavior of the most demanded ceramic pieces for masonry in the region of Norte de Santander, where ceramic blocks represent 15% of the total products manufactured by clay companies [4,5], its demand is related to its constructive functionality and low cost in the market, which makes it an accessible piece and, therefore, a molding element of the thermal loads of the local envelope, a particularly relevant factor in a context of warm semi-arid climate [6,7].

From this perspective, the evaluation of heat fluxes on the physical properties of the parts can determine those products that present better thermal advantages for enclosures in a warm climate with average temperatures between 27 °C and 33 °C without large variations throughout the day, a location with low cloud cover and high solar radiation incidence that can reach 796.8 Wh/m².

2. Methodology and materials

The research selects a specific typology of ceramic piece, block type products of measures length: 300 mm, width: 100 mm and height: 200 mm to analyze 4 constructive units with different shapes in the horizontal cavities Figure 1, the H10 block with six rectangular cavities (B-A), Figure 1(a), the H10 block with 8 rectangular cavities (B-B), Figure 1(b), the H10 block with 4 rectangular cavities whit 2 circular cavities (B-C), Figure 1(c), and the H10 block with 8 circular cavities (B-D), Figure 1(d).

In the study, the same thermal conductivity is applied to all the pieces, obtained through the linear transient flow method of 0.407 W/m°C for clay fired at 1000 °C [8-10], and the influence of the internal shapes of the partitions and the area of the air chambers on the thermal behavior of the products is evaluated; traditional pieces for ceramic masonry are taken as a sample, their characteristics are presented in Table 1.

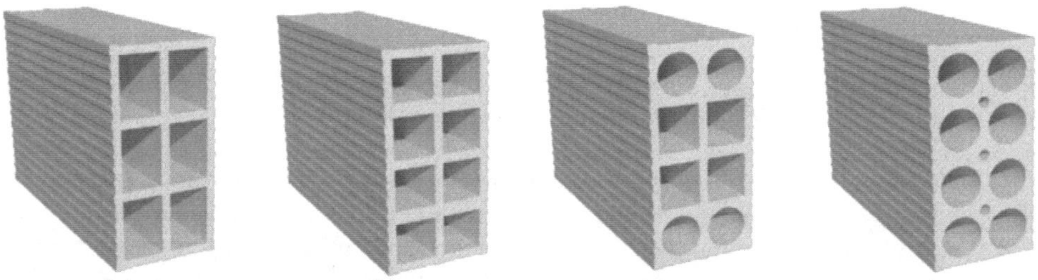
(a) (b) (c) (d)

Figure 1. Ceramic block (a) B-A, (b) B-B, (c) B-C and (d) B-D.

Table 1. Characteristics of the types of ceramic blocks.

Vacuum percentage per air chamber			
B-A	B-B	B-C	B-D
64%	57%	55%	46%

The finite element method is used to develop thermal simulations in ANSYS R16 software, implementing parameters specific to the climate of the city of San José de Cúcuta, Colombia, to establish the thermal behavior by heat transfer of the parts in conditions of 33 °C as average maximum temperature, determining the temperature distribution and heat flows related to the shape and dimension of the parts.

The environmental data used are taken from the IDEAM [11] for a geographic location of latitude: 7.9 °N, longitude: 72.5 °W, altitude: 298 m.a.s.l. in San José de Cúcuta, Colombia, taking as a reference point for the temporality the month of September as the period that presents the highest temperatures throughout the year, considering average maximum climate variables from 12:00 hours to 13:00 hours of a typical day, where, a wind flow presents a speed of 4 m/s and an average maximum solar radiation of 796.8 Wh/ m².

The data used for the conductivity of the materials, $\text{k Clay} = 0.407\ \text{W/m°C}$; in relation to the calculated data, the convective heat transfer coefficient is the value that depends on the wind speed and the temperature and pressure conditions in which it is found, is calculated by the Equation (1) [12].

$$h = \frac{Nu * k}{Lc}, \tag{1}$$

where, h is the convective heat transfer coefficient, Nu is the Nusselt number, k is the thermal conductivity of air, Lc is the assumed characteristic length of 20 cm. The Nusselt number is a dimensionless value that describes the increase in heat transfer over a surface [12]. For a rectangular cross-section and crossflow, is represented by Equation (2).

$$\mathrm{Nu} = 0{,}102\mathrm{Re}^{0{,}675} * \mathrm{Pr}^{1/3}, \quad (2)$$

where Re is the Reynolds number, Pr is the Prandtl number. The Reynolds number is a dimensionless value that describes the behavior of the air flow over the surface of the block and is calculated by the Equation (3) [12].

$$\mathrm{Re} = \frac{\rho * \mathrm{V} * \mathrm{Lc}}{\mu}, \quad (3)$$

where ρ is the air density, V is the wind speed, μ is the dynamic viscosity of the air. The properties of air for a temperature of 33 °C are described as, ρ = $1.1526\ \mathrm{Kg/m^3}$, $\mathrm{k} = 0.026102\ \mathrm{W/m°C}$, $\mu = 0.000018858\ \mathrm{Kg/m.s}$, Pr = 0.72736.

Replacing the values in order of Equation (1), Equation (2) and Equation (3), results in a convective heat transfer coefficient of h = 17.5154 $W/m^2.°C$ to be applied to the outer section of the geometry, where the wind speed takes effect, and assumes a heat transfer by natural convection of 5 $W/m^{2}°C$ and a heat flux of 796.8 W/m^2, for surfaces that are not enclosed as internal air chambers, a heat transfer coefficient by natural convection of h = 5 $W/m^{2}°C$ [12]; the conditions to which the products are subjected are shown in Figure 2.

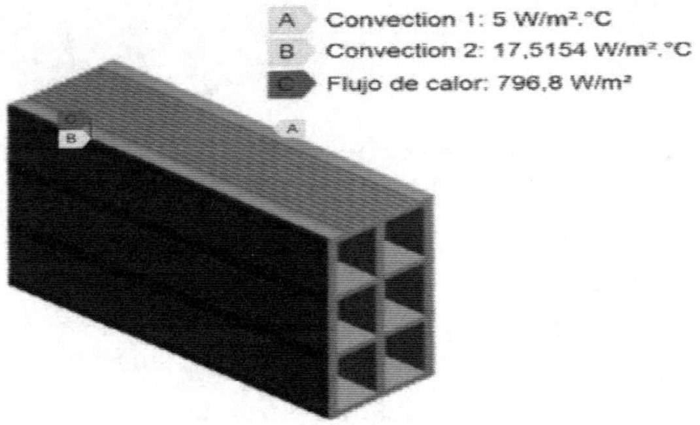

Figure 2. Conditions applied to the models.

3. Results and discussion

The results of the thermal simulations allow a comparative analysis of the heat fluxes and temperature distribution profiles across the bridges and partitions of each piece in relation to the influence exerted by the rectangular and circular geometric shapes of the horizontal air chambers.

3.1. Heat flow results

The Figure 3 shows the results of heat flow profiles in each piece, where the horizontal partitions function as direct thermal bridges, piece B-A (Figure 3(a)) contains 4 direct thermal bridges and B-B (Figure 3(b)), B-C (Figure 3(c)), and B-D (Figure 3(d)) 5 direct heat paths, where, B-C (Figure 3(c)) and B-D (Figure 3(d)) by using a circular geometry increases the percentage of ceramic mass of the thermal bridge and facilitates heat transfer by conduction.

Comparing the shape of the partitions in the rectangular geometries an average heat flow of 149.0 W/m^2 is found and in the circular shapes there is an average of 174.1 W/m^2 with little significant differences. However, if the heat flow in the final wall surface of the pieces is compared higher percentages are observed in B-C (Figure 3(c)) and B-D (Figure 3(d)) as they present less contact with the air chamber, which increases the final heat transferred in the pieces with average values of 100 W/m^2, while geometries such as B-A (Figure 3(a)) and B-B (Figure 3(b)) with greater area exposed to the air chamber decreases the thermal transfer presenting heat flows in the final surface closer to 30 W/m^2; the heat flow results are presented in Table 2.

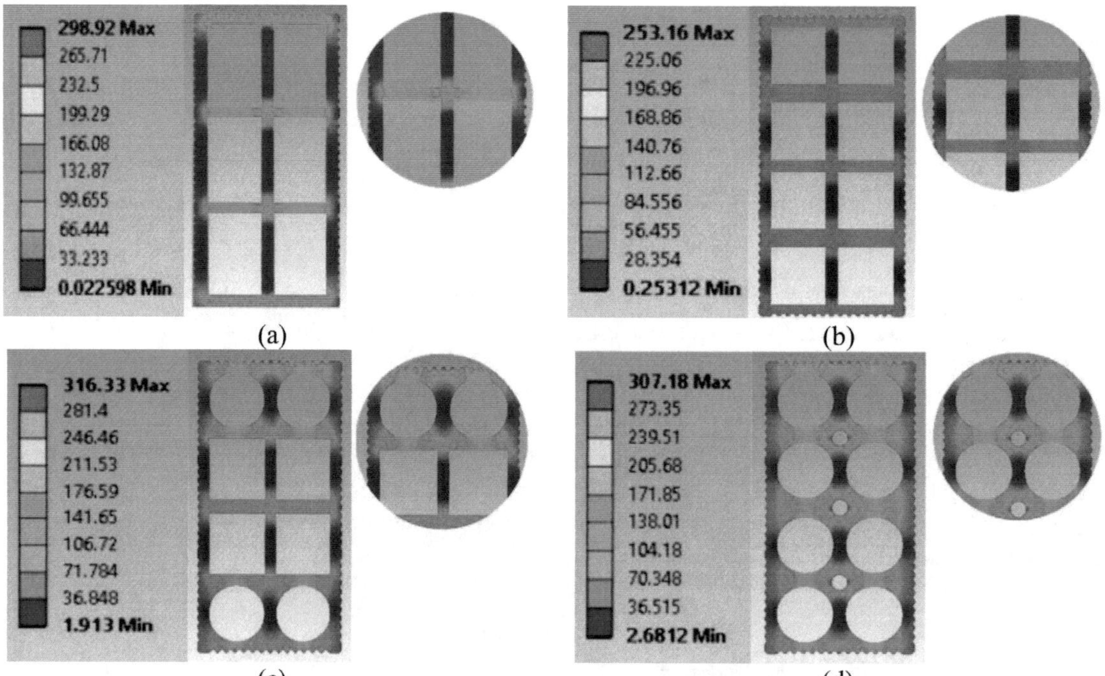

Figure 3. Heat flow results of products type (a) B-A, (b) B-B, (c) B-C and (d) B-D.

Table 2. Heat flow results.

Product	Type	B-A	B-B	B-C	B-D
Average heat flux in partitions (W/m^2)	Rectangular	146.0	140.7	141.6	
	Circular			176.5	171.8

3.2. Temperature distribution results

The temperature transferred to the interior of a B-A Figure 4 product under 33.0 °C conditions present a profile of 37.9 °C, with high temperatures that cross the thermal bridge and reach the internal surface at average temperatures of 41.0 °C; and in those layers that cross air chambers, a transmittance of 38.2 °C is recorded in the final temperatures.

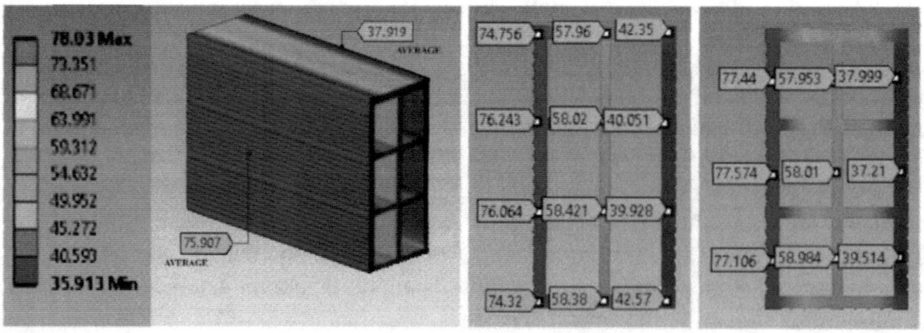

Figure 4. Temperature distribution results for product B-A.

In the Figure 5 the temperature distribution profiles show higher average values in those parts containing circular internal shapes; products such as B-C (Figure 5(b)) containing mixed shapes show temperatures of 38.9 °C and B-D (Figure 5(c)) whose internal cavities are completely curved show temperatures of up to 42.3 °C, product B-B (Figure 5(a)) shows final temperatures of 39.2 °C due to a reduction in the percentages of internal air and an additional thermal bridge with respect to B-A (Figure 4), which shows the lowest final temperature of 37.9 °C, a significant decrease in the final temperature with respect to the other pieces, of approximately 1.0 °C and up to 4.4 °C between B-A (Figure 4) and B-D (Figure 5(c)); the temperature distribution results are presented in Table 3.

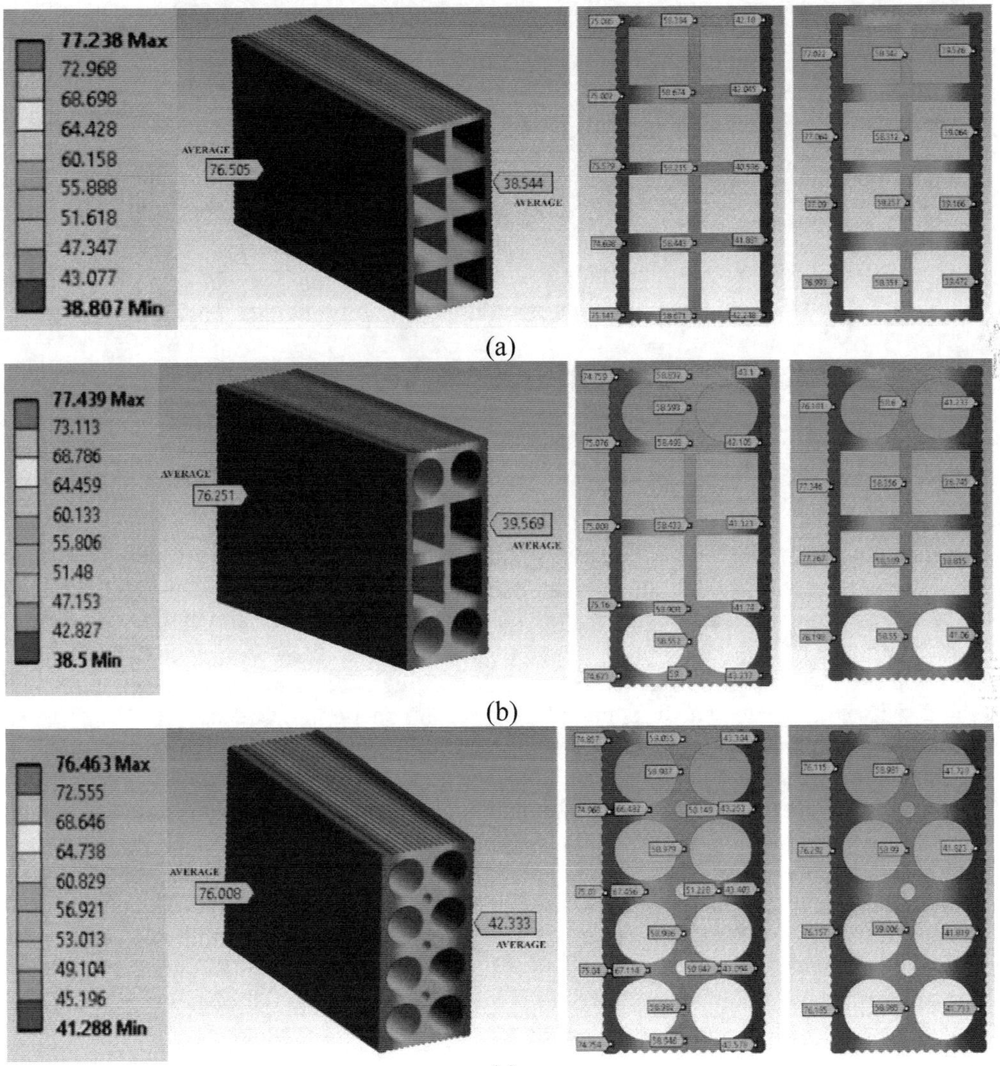

(c)

Figure 5. Temperature distribution results of products type (a) B-B, (b) B-C, and (c) B-D.

Table 3. Temperature distribution results.

Average	B-A	B-B	B-C	B-D
Temperature initial surface (°C)	75.90	76.50	76.25	76.00
Temperature final surface (°C)	37.91	38.5	39.5	42.33

With the results compiled in Table 4, it can be observed that the B-C and B-D pieces that contain cylindrical cavities can present temperatures between 1° C and 2 °C above a B-C piece in the final thermal conditions of the partitions' path, and if the paths through the initial wall, the air chamber, the intermediate wall, the air chamber and the final temperature of the internal wall are analyzed, the results can be from 2.7 °C to 5.0 °C above a B-A type piece. These results allow us to identify the B-A type as the one that, due to its physical characteristics of rectangular cavities and fewer thermal bridges, achieves better thermal performance results for a hot tropical climate.

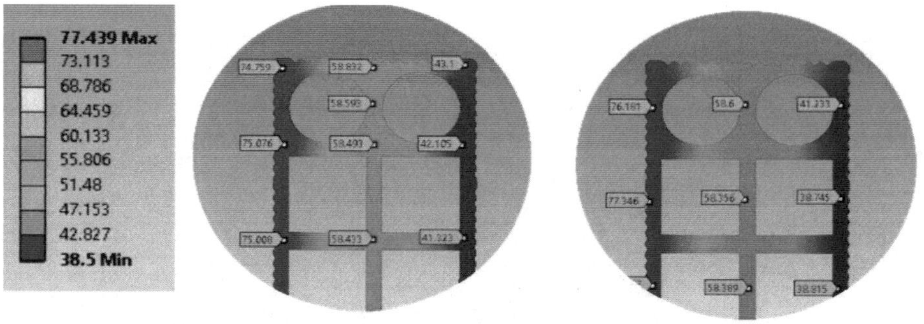

Figure 6. Detail of the results of temperature distribution in part B-C.

Table 4. Final surface wall temperature distribution results.

Average	B-A	B-B	B-C	B-D
Across thermal bridge or partition wall (°C)	41.0	41.7	42.2	43.2
Through wall and air chamber (°C)	36.5	39.2	39.9	41.7

4. Conclusions

One aspect that may be of interest in the selection of materials for the construction of masonry in the area of influence of the ceramic companies in the Norte de Santander, Colombia, is the thermal behavior of the products that build these enclosures. This research has determined the influence of the internal physical characteristics of the ceramic block type pieces on the heat transferred to the final surfaces of the envelope, finding a difference greater than 2 °C with the modification of the number of thermal bridges, the density of the piece and a rectangular or circular shape in its cavities.

The results of thermal simulations demonstrated that B-A, a light piece with 6 rectangular cavities is the most viable product within the available options for constructions in hot semi-arid climates by presenting only three direct thermal bridges, a higher percentage of interior air volume and a rectangular shape. On the other hand, circular shapes and larger mass have shown to obtain temperatures up to 4 °C higher than a B-A block, making it an inefficient system for the envelope in the city of San José de Cúcuta, Colombia.

This comparative study can serve as a basis for the correct selection of building envelope materials in an extreme hot climate, as well as for the definition of design guidelines in the development of new products that can effectively contribute to the reduction of thermal loads of ceramic building systems, a particularly relevant issue in the mitigation of energy consumption in buildings.

References

[1] Sánchez J, Ramírez P 2013 *El Clúster de la Cerámica del Área Metropolitana de Cúcuta* (Cúcuta: Universidad Francisco de Paula Santander)

[2] Sánchez J, González J, Avendaño W 2019 *El Clúster Cerámico: Apuesta de Desarrollo Socioeconómico de Norte de Santander* (Bogotá: Ecoe Ediciones)

[3] Manrique R, Vásquez D, Vallejo G, Chejne F, Amell A, Herrera B 2018 Analysis of barriers to the implementation of energy efficiency actions in the production of ceramics in Colombia *Energy* **143** 575

[4] Gelves J, Monroy R, Sánchez J, Ramírez R 2013 Estudio comparativo de las técnicas de extrusión y prensado como procesos de conformado de productos cerámicos de construcción en el área metropolitana de Cúcuta *Boletín de la Sociedad Española de Cerámica y Vidrio* **52** 1
[5] Rozo S, Sánchez J, Álvarez D 2014 Propiedades físico mecánicas de bloques H10 fabricados en el área Metropolitana de Cúcuta *Ciencia e Ingeniería Neogranadina* **24** 1
[6] Sierra J, Boschmonart J, Gabarrell X 2016 Environmental assessment of façade-building systems and thermal insulation materials for different climatic conditions *Journal of Cleaner Production* **113** 102
[7] Adamczyk J, Dylewski R 2017 The impact of thermal insulation investments on sustainability in the construction sector *Renewable and Sustainable Energy Reviews* **80** 421
[8] Sánchez J, Corpas F, Álvarez Diana 2019 *Aplicaciones de los Nutrientes Tecnológicos en la Industria Cerámica del Área Metropolitana de Cúcuta* (Bogotá: Ecoe Ediciones)
[9] Peña G, Peña J, Gómez M 2014 Determinación experimental de la conductividad térmica efectiva en bloques extinguidos de arcilla roja *Revista Ciencia en Desarrollo* **5** 1
[10] Alvarez D, Sanchez J, Corpas j, Gelves J 2018 Caracteristicas de las materias primas usadas por las empresas del sector cerámico del área metropolitana de Cúcuta Colombia *Boletín de la Sociedad Española de Cerámica y Vidrio* **52** 1
[11] Instituto de Hidrología, Meteorología y Estudios Ambientales (IDEAM) 2020 *Informe Estación Universidad Francisco de Paula Santander, Promedio Horario de Radiación* (Bogotá: Instituto de Hidrología, Meteorología y Estudios Ambientales)
[12] Cengel Y A 2007 *Transferencia de Calor y Masa, Un Enfoque Práctico* (México: McGraw-Hill Interamericana)

Journal of Physics: Conference Series **2118** (2021) 012008 doi:10.1088/1742-6596/2118/1/012008

Design of a mechanical pedal valve to improve the use of water in plumbing systems

L F López Uribe[1], C S Sánchez Rincón[1], M Díaz Jiménez[1], and E Flórez Solano[1]
[1] Grupo de Investigación en Ingenierías Aplicadas para la Innovación, Gestión y Desarrollo, Universidad Francisco de Paula Santander, Seccional Ocaña, Colombia

E-mail: lflopezu@ufpso.edu.co, cssanchezri@ufpso.edu.co

Abstract. The use of water is a very important economic and environmental factor in all industrial sectors since the improvement in its use offers an optimization of this resource. The waste of water usually causes damages firstly to the environment by the supply sources, then to the industries by its high demand and to the hydrosanitary sources that the houses have. The valves have some control that allows a more comfortable use of this resource in sinks that provides some savings. Currently, there are valves that usually replace the traditional ones due to their automatic and/or mechanical operation, which allows for innovation and savings of this resource. In this sense, this research is oriented to the design of a mechanical pedal valve that allows a more comfortable and controlled control, thus granting the operator control in an immediate way, without wasting the fluid in the middle of the washing and avoiding a manual contact that in a certain way dissipates the risk of contagion of diseases such as Covid-19 by direct manipulation. An analytical methodology supported by computer aided drafting programs such as Ansys Fluent and SolidWorks was used. Finally, it was determined that, for the dynamic and static studies, the designed valve complies with the Colombian technical standard and the static design factors.

1. Introduction

Hydraulic valves have brought a great development to hydraulic systems in most industrial applications and in homes with plumbing systems, since their use is not only based on the control of water flow, but also on the way in which this valuable resource can be saved. A large waste of water leads to large environmental, economic and energy losses. Therefore, many authors have developed research oriented to the innovation of these, in order to take them to another level of production and optimization [1,2]. On the other hand, the field of computational fluid dynamics (CFD) is a methodology of great help for the study of the behavior of the variables involved in a hydraulic system, these variables are the velocity and pressure of the liquid stream flowing around or inside the object under study [3-7]. In this way, the use of specialized software for the study of these dynamic processes of fluids and static processes of valve design becomes more acute, since continuous and discrete particle systems are easily simulated by means of these tools such as MATLAB® and Ansys, since their programming language presents a high complexity of equations for the analysis providing a low percentage of error.

Factors such as the inadequate use of traditional domestic valves have caused a disproportionate loss of this resource in the basic manual washing process, since the faucet is usually kept open while the operation is being completed. In view of the situation posed by the need for a more controlled use, this research is based on the design of a mechanical pedal valve, where the operator has the facility to complete the task with a moderate use of the surrounding fluid, thus avoiding waste, and providing comfort in the process. For this, a descriptive methodology was used which was based on the static and dynamic analysis of the valve supported by SolidWorks and Ansys Fluent, which took as design support the parameters imposed by the "Norma Técnica Colombiana (NTC), NTC 1500" [8]. Finally, this design represents a positive impact in the adequate use of water and in the reliability of its operation under real and ideal conditions.

2. Methodology

This is a descriptive type of research with a quantitative approach based on different calculations and measurements that allowed structuring the design of a spring-returned mechanical pedal valve as an alternative to improve the use of water through faucets in plumbing systems. In a pedal valve the user controls with the foot the flow of fluid through a pedal actuator that pushes a stem allowing the circulation of water through a pipeline, for this purpose the loads and pressures that interact in the body of the valve, the stem and the pedal must be analyzed. Therefore, these parts are studied as a discrete system of particles, which is related to physical principles such as Newton's second law with the development of the equations to perform the relevant static and dynamic analysis [9-11]. In addition, this process was developed in the environment of computer aided design (CAD) software that is based on finite element analysis (FEM).

2.1. Model of mechanical elements

Next, the modeling of the mechanical elements of the system was developed by the SolidWorks® software that presides high precision by the complexity of the programming language that supports it. Figure 1 shows the parts that make up the established design, such as the valve body, cover, spring, stem, pedal base, and pedal; then Table 1 shows the dimensions of the critical parts of the valve.

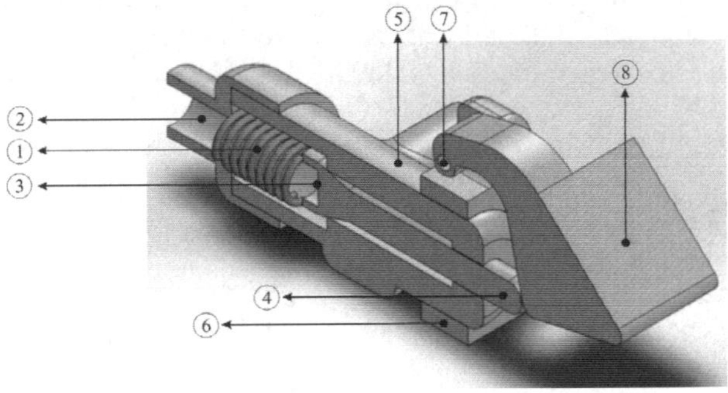

Figure 1. Pedal valve parts.

Table 1. Parts and critical areas of concentration.

	Piece	Area (mm^2)
(1)	Spring	– – –
(2)	Valve body inner conduit	7642
(3)	Internal surface of the rod	81.68
(4)	External surface of the rod	132.11
(5)	Body cover	– – –
(6)	Pedal base	– – –
(7)	Valve joint	– – –
(8)	Pedal face	2243.26

2.2. Simulation of fluid behavior through the valve

The geometric design of this valve was subjected to a fluid analysis with the Ansys fluent tool to observe the behavior of the pressures and velocities in conditions of daily use established in the range of the NTC 1500 [8] varying the obstruction of the water flow. For this purpose, the hydraulic duct of a dishwasher system used in most homes was represented as an example, in which the conventional handles were replaced by the proposed design and results were obtained from the idealized evaluation in the software; K-EPSILON standard was the evaluation model used in the simulation. The kinetic energy of turbulence, k, and its dissipation rate, ϵ, are obtained from the following transport Equation (1) and Equation (2) [12].

$$\frac{\partial}{\partial t}(\rho k) + \frac{\partial}{\partial x_i}(\rho k u_i) = \frac{\partial}{\partial x_j}\left[\left(\mu + \frac{\mu_t}{\sigma_k}\right)\frac{\partial k}{\partial x_j}\right] + G_k + G_b - \rho\epsilon - Y_M + S_k, \quad (1)$$

$$\frac{\partial}{\partial t}(\rho\epsilon) + \frac{\partial}{\partial x_i}(\rho\epsilon u_i) = \frac{\partial}{\partial x_j}\left[\left(\mu + \frac{\mu_t}{\sigma_\epsilon}\right)\frac{\partial \epsilon}{\partial x_j}\right] + C_{1\epsilon}\frac{\epsilon}{k}(G_k + C_{3\epsilon}G_b) - C_{2\epsilon}\rho\frac{\epsilon^2}{k} + S_\epsilon, \quad (2)$$

where, G_k represents the turbulence kinetic energy generation due to mean velocity gradients, G_b is the turbulence kinetic energy generation due to buoyancy, Y_M represents the contribution of fluctuating dilatation in compressible turbulence to the overall dissipation rate, $C_{1\epsilon}$, $C_{2\epsilon}$, and $C_{3\epsilon}$ are constants. σ_k, σ_ϵ are the turbulent Prandtl numbers for k and ϵ, respectively. ρ is the fluid density, u_i is the flow velocity, x_i, x_j are the positions, S_k and S_ϵ are user-defined source terms and μ_t is the turbulent viscosity.

This method uses standard wall functions based on the analysis of the boundary layer that the fluid has with the internal walls of the device, point to point operations are performed depending on the mesh to be used. The logarithmic law of mean velocity valid and reliable results must meet in a predetermined range "Yplus" or "Y*" between $30 < Y* < 300$ [13,14]. Table 2 shows the parameters and variables taken for the simulation, such as mesh type and number of elements. According to the standard in its second update, it indicates that the operation of the service of the water fluid for hydro-sanitary devices consists of a minimum pressure of 5 kPa and a minimum flow rate of 0.19 L/s. However, for real conditions of use, an experimental method of measurement of specific pressures was carried out for this research, in homes in different areas of the city of Ocaña, Norte de Santander, Colombia, it was obtained that the highest pressure found in the homes was 242 kPa.

Table 2. Dynamic simulation parameters in Ansys Fluent.

Type of mesh	Tetrahedral with edge tuning
Number of nodes	88880
Number of elements	404775
Pipe diameter	½"
Inlet pressure	242 kPa
Outlet pressure	Atmospheric
Type of fluid	Water (H_2O)
Fluid characteristic	Turbulent (k-epsilon) no-slip condition
Iterations	100

2.3. Static analysis of the valve

Following the dynamic study, a static analysis was performed in SolidWorks by means of FEM in the design against maximum pressure conditions. The simulation was developed considering the valve closed, this implies pressure of 242 kPa in the internal walls, fastenings with screws that fix the device to the floor and a bolt that joins the base of the valve and the pedal. Most valves and faucets are made of metallic and polymeric materials, the most commonly used being copper, steel, polyvinyl chloride (PVC) and high-density polyethylene (HDPE) [15,16]. Based on the above, it was decided to use a material with excellent resistance to traction, compression, low density, chemically inert, high impermeability, and ductility. In this sense, the selection was oriented towards (HDPE).

2.3.1. Meshing. The Jacobian points are used to determine the degree of confidence of the mesh, they verify that the contours of the elements do not intersect, if a Jacobian point is negative the software stops immediately. Table 3 shows the static simulation parameters; another way with which the software makes a quality check on the mesh elements is by means of the aspect ratios, it does this by making sure that the tetrahedral elements are perfect and uniform whose edges are of equal length [17]. Figure 2(a) shows the fixed points for the analysis and in Figure 2(b) the total meshing.

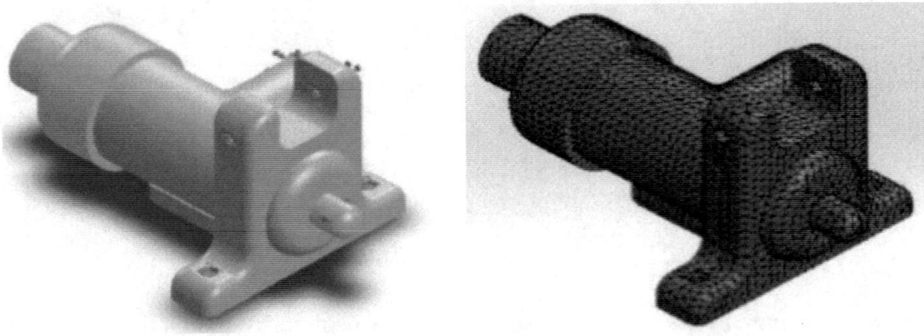

(a) (b)

Figure 2. Meshing process. (a) fixed points established for the static analysis, (b) total valve meshing.

Table 3. Static simulation parameters in SolidWorks.

Type of mesh	Tetrahedral with edge tuning
% Of distorted elements	0
Jacobian points	4
Element size	2.39 mm
% Of elements with aspect ratio > 10	0.02
Total number of nodes	72235
Total number of elements	45012

2.3.2. Von Mises stress and factor of safety. To calculate the element stresses, the program averaged the corresponding nodal stresses for each element. In most cases, the yield stress (which is defined by the selected material) is used as the stress limit. However, SolidWorks allows the use of the tensile/rupture stress limit. The software offers criteria with which the safety factor distribution is calculated, the recommended one for linear static analysis and ductile materials is the Von Mises Max stress criterion. This says that the safety factor (SF) will be equal to the quotient between the yield stress of the material and the Von Mises Max stress, since the material will start to fail when the yield stress and the stress are equal [18]. Concerning the principal stresses σ^1, σ^2, σ^3, the Von Mises stress is expressed by Equation (3); where, σ_e is the maximum Von Mises stress.

$$\sigma_e = \sqrt{\frac{(\sigma_1-\sigma_2)^2 + (\sigma_1-\sigma_3)^2 + (\sigma_2-\sigma_3)^2}{2}}, \quad (3)$$

3. Results

In this section it reports the detailed analysis of the data in two sections governed by the methodological order established from the beginning, in which the analysis of the dynamic behavior of the fluid and the static behavior of the device are specified, respectively.

3.1. Dynamic analysis in the valve

The following graphs illustrated in Figure 3, in conjunction with Table 4 show the result of the variation of pressure and velocity as the stem obstructs the conduit, the condition reports that the fully open stem has moved 0 mm and when closed, its displacement will have been 10 mm as shown in Figure 3. Then, it is observed that in Figures 3(a) and Figure 3(b) the variation of pressure and velocity respectively, does not present considerable fluctuations between 0 mm and 5 mm of opening.

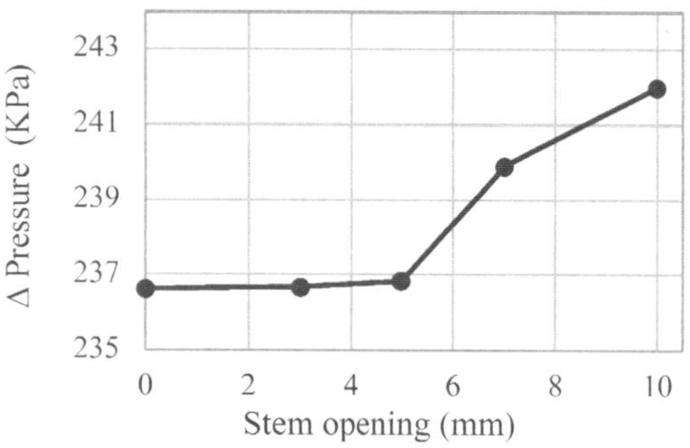

(a)

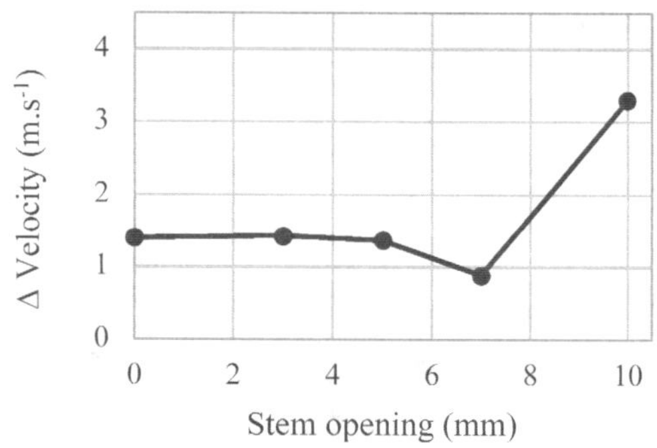

(b)

Figure 3. Representation of variables. (a) pressure drop growth as a function of stem opening, (b) velocity drop growth as a function of stem opening.

Table 4. Simulation results in Ansys Fluent.

Opening in obstruction (mm)	Inlet velocity ($m\ s^{-1}$)	Inlet pressure (kPa)	ΔVelocity ($m\ s^{-1}$)	**Δ**Pressure (kPa)	Output velocity ($m\ s^{-1}$)	Minimum output parameter NTC 1500 [8]	
						Flow rate ($L\ s^{-1}$)	Pressure (kPa)
						0.19	5.00
0	3.10	242.00	1.40	236.62	1.69	0.21	5.38
3	3.10	242.00	1.43	236.67	1.66	0.21	5.33
5	3.10	242.00	1.36	236.84	1.73	0.21	5.16
7	3.10	242.00	0.88	139.91	2.21	0.28	102.09
10	3.10	242.00	3.10	242.00	0.00	0.00	0.00

Likewise, based on the previous data, the minimum pressure at which the fluid will leave the tap and the minimum flow rate at which the service will be obtained were tabulated, thus ensuring that it complies with the parameters of the NTC 1500 [8]. Next, Ansys Fluent evaluated the effectiveness of the simulation with the Yplus parameter. Figure 4 corroborates that the dynamic simulation of the continuous system complies with the established range.

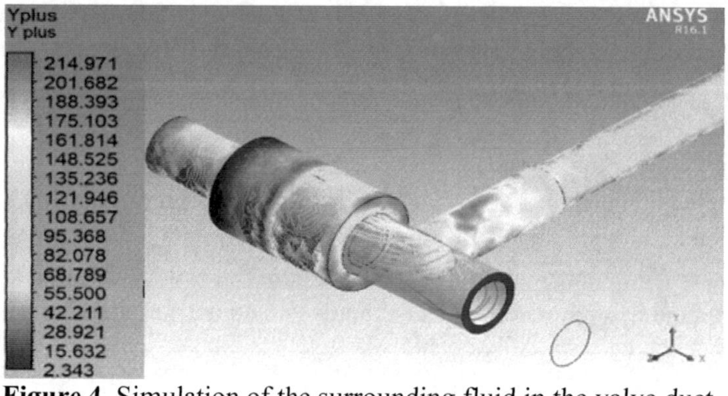

Figure 4. Simulation of the surrounding fluid in the valve duct.

3.2. Static analysis of the valve

The unit deformations were calculated from the displacements, which show that the critical deformations are in the contact zone in the valve body inner conduit. On the other hand, the largest deformation is 2.461×10^{-5} mm and the minimum deformation value, which is distributed in most of the model, is negligible; the static simulations acquired from the isolation of the critical parts showed the effectiveness of the maximum unit deformation, Von Mises Max stress and minimum SF, as shown in Table 5.

Table 5. SolidWorks static simulation summary.

Piece	Max deformation (mm)	Von Mises max effort (Nmm^{-2})	Safety factor min
Stem	0.02	23.67	20.93
Pedal	0.02	2.69 x106	10.58

4. Conclusions

The static study of the maximum deformations under critical working conditions according to the standards shown in this work provided high reliability in the design executed. At the same time, it was shown that the pedal has the lowest safety factor, which is an approved value for the application to be fulfilled in the design. On the other hand, the valve body shows the least uniform distribution of the applied stresses, however, it is congruent to state that the stresses do not generate a significant risk of fracture in the body structure. In the hydraulic analysis, it was demonstrated that the operation of this device fits the minimum parameters stipulated by Colombian technical standards since the minimum outflow of the tap is 0.21 Ls^{-1} and the minimum outlet pressure is 5.16 kPa being compared with the standard: 0.19 Ls^{-1} and 5 kPa respectively. Finally, it is plausible to mention that the use of this kind of device represents a hygienic alternative to the Covid-19 sanitary contingency, due to the low manual interaction in comparison with traditional taps.

Acknowledgments

To the Universidad Francisco de Paula Santander, Ocaña, Colombia, the INGAP research group of the “Programa de Ingeniería Mecánica” and the “División de Investigación y Extensión” for supporting the exploration of new technological advances.

References

[1] Sun X, Kim H S, Yang S D, Kim C K, Yoon J Y 2017 Numerical investigation of the effect of surface roughness on the flow coefficient of an eccentric butterfly valve *J. Mech. Sci. Technol.* **31** 2839

[2] Mu Y, Liu M, Ma Z 2019 Research on the measuring characteristics of a new design butterfly valve flowmeter *Flow Meas. Instrum.* **70** 101651

[3] Aabid A, Khan S A 2021 Investigation of high-speed flow control from CD nozzle using design of experiments and CFD Methods *Arab. J. Sci. Eng.* **46** 2201

[4] Bhuyan D, Kumar K 2017 Computational fluid flow analysis of base valve for twin tube shock absorbers *5th International Conference of Materials Processing and Characterization (ICMPC 2016)* ed Swadesh K (Ranchi: Materialstoday) p 95

[5] Mohamed A, Kchaou H, Abid M S, Driss Z 2018 Impact of stern design on hydrodynamic drag of auv’s hull *Indian J. Geo-Marine Sci.* **47** 89

[6] Dong X, Zhang Z, Liu D, Tian Z, Chen G 2018 Numerical investigation of the effect of grids and turbulence models on critical heat flux in a vertical pipe *Front. Energy Res.* **6** 58:1

[7] White F M 2011 *Introduction to Fluid Mechanics, 8th edition* (New York: McGraw-Hill)

[8] Instituto Colombiano de Normas Técnicas y Certificación (ICONTEC) 2004 *Código Colombiano de Fontanería, Norma Técnica Colombiana, NTC 1500* (Colombia: Instituto Colombiano de Normas Técnicas y Certificación)

[9] Cekus D, Gnatowska R, Kwiatoń P, Šofer M 2019 Simulation research of a wind turbine using SolidWorks software *J. Phys. Conf. Ser.* **1398** 012001:1

[10] Zhou X M, Wang Z K, Zhang Y F 2017 A simple method for high-precision evaluation of valve flow coefficient by computational fluid dynamics simulation *Adv. Mech. Eng.* **9** 1

[11] Li Z, Zhu H, Zhang J 2020 Design and online calibration methods of pressure-independent intelligent regulating valve based on hydrodynamic resistance characteristics *Energy Build.* **224** 110227

[12] Ansys, Inc 2009 *Ansys* (Pensilvania: ANSYS, Inc)

[13] Xu Y, Jiang F, Chen J, Li S 2018 Design and optimization of the faucet flow channel structure *Mech. Mach. Sci.* **55** 595

[14] Zou Y, Zhao X, Chen Q 2018 Comparison of STAR-CCM+ and Ansys Fluent for simulating indoor airflows *Build. Simul.* **11** 165

[15] Fan J, Njuguna J 2016 *An Introduction to Lightweight Composite Materials and Their Use in Transport Structures* (Cambrige: Woodhead Publishing)

[16] Mahakul R, Nath Thatoi D, Choudhury S, Patnaik P 2019 Design and numerical analysis of spur gear using SolidWorks simulation technique *Mater International Conference on Recent Advances in Mechanical Engineering Research and Development (ICRAMERD-20)* ed Seshadev S (Odisha: Materialstoday) p 145

[17] González O, González C, López A 2019 Introducción a análisis de elemento finito: SolidWorks y MATLAB *Ideas en Ciencias la Ing.* **1** 28

[18] Mott R L, Vavrek E M, Wang J 2018 *Machine Elements in Mechanical Design, 6th edition* (New York: Pearson Prentice Hall)

Fatigue in asphalt mixtures – a summary to understand the complexity of its mathematical modeling

H A Rondón-Quintana[1], F A Reyes-Lizcano[2], and C A Zafra-Mejía[1]
[1] Facultad del Medio Ambiente y Recursos Naturales, Universidad Distrital Francisco José de Caldas, Bogotá, Colombia
[2] Facultad de Ingeniería, Pontificia Universidad Javeriana, Bogotá, Colombia

E-mail: harondonq@udistrital.edu.co, fredy.reyes@javeriana.edu.co

Abstract. Based on the reviewed literature in relation to the phenomenon of fatigue in asphalt mixtures, the foregoing paper depicts and describes in summary, the main variables that impact in the generation of said phenomenon in asphalt pavements. This has the purpose of showing its complexity to mathematically model it. As a general conclusion obtained in the study, it was found that the calibration difficulty of the models is mainly since the mathematical equations must be in capacity of considering that fatigue resistance of asphalt mixtures depends on load mode (stress-controlled or strain- controlled), the type of load (haversine or sinusoidal) and the rest periods to which laboratory samples are subjected. Additionally, both in situ, as within the laboratory, this varies with stiffness, volumetric composition (type and content of asphalt and aggregate), the geometry of samples, with effects associated to mix durability and environmental conditions, with the type of test, border conditions and support layers (base, subbase, subgrade). If these physical parameters are not considered, the mathematical equations lose reliability.

1. Introduction

One of the main damage mechanisms for asphalt mixtures is fatigue cracking [1-8]; the continuous passing of vehicles over asphalt pavements makes the asphalt layer flex, generating tension stresses in its bottom layer. This load repetition makes the asphalt mix lose stiffness and this produces the appearing of tensile strains that are non-recoverable, which in turn lead to the forming of microcracks. The continuous load repetitions and the reduction of stiffness, causes coalescence of microcracks that leads to the forming of a visible fissure (macro-cracks) on the surface. Normally, these are seen as parallel cracks directed on a longitudinal sense of the roadway which are directly under the application of load (under the tires).

Understanding the manner in which this damage mechanism occurs in an asphalt pavement is a complex task. This is because there are multiple physical variables that impact in its generation [7]. Additionally, most studies on the subject are carried out within the laboratory and extrapolating these results with those obtained on a real scale, increases the level of complexity. Despite the broad number of research conducted on the subject, there is not yet an existing definitive model that characterizes and predicts fatigue resistance for asphalt mixtures in service.

This paper presents and describes in summary, the most significant physical variables that impact in the generation of this damage mechanism. Said variables are presented with the purpose of showing the difficulty for obtaining a prediction mathematical model for the phenomenon of fatigue in asphalt mixtures. Additionally, the information presented can serve as a consulting tool for researchers that develop constitutive equations and regression models associated to the phenomenon of fatigue in asphalt mixtures. Not including all physical variables that influence the fatigue resistance of asphalt pavements could results in equations with low reliability.

2. Factors affecting fatigue response

The most important factors influencing the fatigue life of asphalt pavements are described below.

2.1. Load

Within the laboratory, when fatigue resistance tests are carried out, samples are generally subjected to two types of loads: stress-controlled (σ-C) and strain-controlled (ε-C). In σ-C tests, while the stress that is applied on the sample is kept constant during the whole test, strain increases [9]. In ε-C tests, while sample strain is constant throughout the entire test, stress is reduced. In other words, under ε-C, with the increase of load repetitions, the asphalt mixture deteriorates, losing stiffness, and therefore it requires less load and less stress to produce the same strain [10].

Some factors associated to the type of load that add difficulty in obtaining models that predict the fatigue life process of asphalt mixtures are the following: (i) the response of asphalt mixtures in both types of loads (σ-C and ε-C) tends to be opposite (see Table 1). Generally, mixes tend to display a higher fatigue resistance under σ-C when their stiffness increases. The opposite case occurs the load mode is ε-C [6,11-13]; (ii) for laboratory tests to simulate what happens on the field as closely as possible, it is necessary to know the load mode under which the asphalt mixture works when in service. According to [14], thin asphalt layers (with thickness equal or under 2 inches) are put under ε-C, while those with a thickness over 6 inches are put under σ-C. Thickness layers that are over 2 inches but under 6 inches are loaded in an intermediate mode; (iii) failure criteria applied to both types of loads are different [9]. Generally, under ε-C, the sample is supposed to fail when it has lost half of its initial stiffness, while under σ-C, it is assumed that failure occurs when the initial deformation is duplicated or when there is total rupture of the sample [15,16]; (iv) according to several authors, tests under ε-C present more disperse results than tests carried out under σ-C [9-11,17]; (v) loading and unloading processes create dissipated energy (area of the stress-strain curve) [18]. This energy increases under σ-C and decreases under ε-C [19].

Table 1. Comparison of Tests under σ-C and ε-C [20].

Description	σ-C	ε-C
Usual failure criterion	Sample rupture	Loss of one half of initial stiffness
Fatigue life	Shorter	Longer
Results dispersion	Lower	Higher
Temperature increase	Reduction of life	Increase of life
Stiffness increase	Increase of life	Reduction of life
Rest time period	increase of life	reduction of life

On the other hand, on the field, the application of load is different to the one used in laboratory tests. The type, magnitude, frequency, and rest periods between load applications are different. Generally, laboratory fatigue tests are done by applying a continuous load. However, discontinuous tests, or tests that present rest periods between the application of load, better represent in situ conditions [3]. The recovery of material, as a product of its restoring between load applications, results in a partial increase of stiffness after this value has diminished, and fatigue life is prolonged [21]. This increase in fatigue life as a product of greater rest periods is greater as asphalt content and temperature increase [22,23]. It is important to understand the beneficial effects of rest periods are mainly due to the closing of microcracks within the asphalt mixture [24,25].

According to [26], self-healing occurring during rest periods has been shown to increase the fatigue life of asphalt mixes. With regards to frequency, it is generally reported that it slightly influences on fatigue test results [27,28]. Additionally, in laboratory fatigue tests, loads can be type sinusoidal and haversine waveforms. The responses of asphalt mixtures in both types of loads are different [29,30], increasing the complexity level for modelling. Equation (1) and Equation (2) relate number of cycles up to complete failure (N_f) with the tensile strain (ε_t) or tensile stress (σ_t) expressed through Wöhler curves [31-33].

$$N_f = k_1 \left(\frac{1}{\varepsilon_t}\right)^{k_2}, \quad (1)$$

$$N_f = k_1 \left(\frac{1}{\sigma_t}\right)^{k_2}. \quad (2)$$

k_1 y k_2 obtained through regression are difficult to predict, given that they vary based on multiple factors which will be introduced as follows.

2.2. Stiffness

One of the mechanical parameters that is most used in equations that attempt to simulate the response of asphalt mixtures to cracking due to fatigue is stiffness (E). A very stiff asphalt mixture in situ may not display a good behavior towards fatigue when the asphalt layer is thin, however, in thick asphalt layers the opposite may occur. This response is not completely simulated in the laboratory and because of this reason, it is difficult to mathematically model it. Now, this degree of difficulty increases if we consider that that the asphalt mixture stiffness is a function of multiple factors such as asphalt binder stiffness, asphalt binder content and type (asphalt cement, asphalt emulsion, cutback asphalt, foamed asphalt, modified asphalt binders, etc.) [34], form, texture and absorption of the aggregate [35,36], filler type and content [37,38], volumetric composition of the mix (air void contents (Va), voids filled with asphalt (VFA), among others) [39], temperature, load frequency, traffic and environment [40], chemical composition, the aging of the asphalt [41], among others. Some of these factors cause said mechanical parameter in situ changes across time (climate variation, aging and chemical composition) further hindering its modeling.

In general, mixes with greater Va display shorter fatigue life [42-46]. On the other hand, fatigue response is a combination between the aggregates gradation, Va and the asphalt content in volume (Vb) [47]. According to [42,44,48], there is an optimal asphalt content (OAC) that insures the best behavior of the asphalt mixture towards fatigue. Apparently, this OAC could correspond to that which is required for attaining maximum stiffness in the mix and is superior to the OAC obtained based on the Marshall method [3]. However, on dense mixes, within certain ranges of Va and Vb, the main criterion that could control the response of mixes towards fatigue is the type of load [49]. This is given because in this type of mixes, less Va and Vb tend to increase stiffness, and under σ-C, fatigue life could increase, while under ε-C, the opposite could occur.

The same happens in terms of asphalt temperature and stiffness; when temperature is lowered, asphalt stiffness increases, increasing stiffness in the mix. Therefore, under σ-C tests, mixes that have more stiff asphalt binders tend to present a greater fatigue life (opposite case occurs in tests under ε-C [10]). On the other hand, in general, mixes that use modified asphalt (using polymers, fibers and nanomaterial, among others) tend to display greater fatigue life [14,50]. Depending on the type of modifier and its content, the response to fatigue in mixes changes. This makes it difficult to obtain regression parameters in Equation (1) and Equation (2).

Asphalt mixtures in service age and increase their stiffness across time. Additionally, asphalt goes from a ductile to a brittle behavior when it ages; apparently aging also tends to be related to the mode of load. Under σ-C, aged mixes tend to display longer fatigue life [51], while under ε-C the opposite occurs [49]. Therefore, under low temperatures, stiffness increases in asphalt due to aging reduces the mix's resistance to failure, due to the increase of brittle, especially in thin asphalt layers [42,52].

On the other hand, depending on the compaction method of laboratory samples (static, impact or Marshall compaction, vibratory, kneading, gyratory and with wheels and rollers), the response in fatigue tests is different [53], given that samples present variation in the orientation and distribution of aggregates, and the distribution of voids [17,54], which directly influences in its stiffness. Equation (3) to Equation (10) correlate N_f with E and volumetric composition of mixes [55-65]. These equations also have the main complexity of predicting regression parameters k_i, β_{fi}, which broadly vary with the above-mentioned physical factors. E*: dynamic modulus (psi); ϕ: phase angle.

$$N_f = k_1 \left(\frac{1}{\varepsilon_t}\right)^{k_2} \left(\frac{1}{E}\right)^{k_3} \frac{V_b}{Va+V_b}, \tag{3}$$

$$\log N_f = k_1 - k_2 \times \log \varepsilon - k_3 \times \log E, \tag{4}$$

$$N_f = k_1 \left(\frac{1}{E}\right)^{k_2}, \tag{5}$$

$$N_f = 18.4 \times 10^{M} (4.325 \times 10^{-3} \varepsilon_t^{-3.291}) |E *|^{0.854}, \tag{6}$$

$$M = 4.84 \left(\frac{V_b}{V_b+V_v} - 0.69\right), \tag{7}$$

$$N_f = k_1 (k_2 V_b + 1.08)^{k_3} \left(\frac{1}{\varepsilon_t}\right)^{k_3} \left(\frac{1}{E}\right)^{k_4}, \tag{8}$$

$$N_f = k_1 \exp(k_2 VFA)\, {\varepsilon_o}^{k_3} (E \sin \varphi)^{k_4}, \tag{9}$$

$$N_f = \beta_{f1} k_1 \varepsilon_t^{-\beta_{f2} k_2} E^{-\beta_{f3} k_3}. \tag{10}$$

2.3. *Test type, border conditions and support layers stiffness*

Another aspect that makes the mathematical modeling and prediction of fatigue life of asphalt mixtures difficult is the test type (tension-compression, 2,3,4 points bending, indirect tensile, dynamic shear, etc.), the geometry of samples (cylindrical, beams, trapezoidal samples, etc.), border conditions (lateral confinement, thickness, topography, etc.) and support layers (base, subbase, subgrade) stiffness [20,66]. These factors cause results of laboratory tests to be different. Because of this reason, several researchers have found broad discrepancies between the deformations measured in laboratory and in situ and have determined some correlation coefficients that are known in the reference literature as "shift factor". The problem of these coefficients is that they are disperse (they fluctuate between 13 and 700) [59-60].

3. Conclusions

This study shows the main physical variables that impact on the phenomenon of fatigue in asphalt mixtures under repeated loads, with the purpose of considering them in future studies; considering the bibliographical review carried out, it is concluded that difficulty in mathematically modelling the phenomenon of fatigue in asphalt mixtures is mainly given because: fatigue life in asphalt mixtures depends on the mode of load (σ-C or ε-C) under which lab samples are subjected to. In both types of loads, the response in mixes tends to be opposite. In turn, the most appropriate mode of load for using in laboratory tests is in function of asphalt layer thickness in pavement. Fatigue life in asphalt mixtures depends on the rest time between load applications and the type of wave (haversine or sinusoidal) used for tests.

Traditionally, fatigue life in asphalt mixtures is correlated to the form in which its stiffness evolves. Therefore, factors such as voids and asphalt content, compaction degree, type of aggregate and asphalt, form and size of particles, aging, load frequency and temperature are analyzed from the standpoint of how stiffness evolves, in order to then evaluate the increase or decrease of fatigue life. Considering all that was mentioned, if each of these factors increases mix stiffness, this material will tend to develop an increase in its fatigue life when it is subjected to σ-C. An opposite response could occur for the case of mixes subjected to the load mode of ε-C.

Few equations in the world simulate aspects associated to durability (climate effect, aging, among others) and the compaction processes of asphalt mixtures (in laboratory and in situ). The mathematical equations used for attempting to predict the response towards fatigue in situ and laboratory are generally focused in mainly relating the number of failure cycles that the mix resists with applied tensile strain or stress, with its stiffness and volumetric composition. The greatest deficiency of these mathematical relations is that they are unable to predict in a single equation, strain and/or number of failure cycles that an asphalt mixture will undergo when subjected to different magnitudes of load or deformation, stiffness, and environmental conditions. Likewise, these are uncapable of considering, within a single mathematical equation, aspects associated to test type, sample geometry, border conditions and support layers stiffness.

Mathematical equations must take in account all physical variables that influence the fatigue resistance of asphalt pavements to predict it in a reliable manner. Otherwise, pavement engineers could make inaccuracies when designing asphalt pavements. Future research should further develop and design an experimental lab phase that attempts to simulate all the physical variables identified in this review.

References

[1] Ahmed T M, Al-Khalid H, Thamer J 2019 Review of techniques, approaches and criteria of hot-mix asphalt fatigue *Journal of Materials in Civil Engineering* **31(12)** 03119004

[2] Lv S, Hu L, Xia C, Wang X, Borges-Cabrera M, Guo S, Chen J 2020 Development of fatigue damage model of asphalt mixtures based on small-scale accelerate pavement test *Construction and Building Materials* **260** 119930

[3] Hsu T W, Tseng K H 1996 Effect of rest periods on fatigue response of asphalt concrete mixtures *Journal of Transportation Engineering* **122(4)** 316

[4] Martono W, Bahia H U, D'Angelo J 2007 Effect of testing geometry on measuring fatigue of asphalt binders and mastics *Journal of Materials in Civil Engineering* **19(9)** 746

[5] Masad E, *et al.* 2008 A unified method for the analysis of controlled-strain and controlled-stress fatigue testing *International Journal of Pavement Engineering* **9(4)** 233

[6] Tarefder R A, Kias E, Zaman A 2008 Cracking in asphalt concrete under wet and dry conditions *Pavements and Materials 2008: Modeling, Testing, and Performance, Proceeding of the Symposium on Pavement Mechanics and Materials* (Minneapolis: ASCE press) p 37

[7] Nian T, Ge J, Li P, Guo R, Liu W 2020 Influence of multiple factors on the shear fatigue resistance of asphalt pavement interlayer adhesive materials *Journal of Materials in Civil Engineering* **32(9)** 04020251

[8] Jin G-L, Zhang Z-X, Huang X-M 2021 Fatigue cracking analysis of asphalt concrete based on coupled XFEM-continuum damage mechanics method *Journal of Materials in Civil Engineering* **33(1)** 04020425

[9] Di Benedetto H, De La Roche C, Baaj H, Pronk A, Lundström R 2003 Fatigue of bituminous mixtures: different approaches and RILEM group contribution *Proceedings of the 6th International RILEM Symposium* (Held Zurich: RILEM Publications S.A.R.L) p 15

[10] Epps J A, Monismith C L 1972 *Fatigue of Asphalt Concrete Mixtures–Summary of Existing Information Fatigue of Compacted Bituminous Aggregate Mixtures* ed Gallaway B (West Conshohocken, PA: ASTM STP 508) p 19

[11] Di Benedetto H, De La Roche C, Baaj H, Pronk A, Lundström R 2004 Fatigue of bituminous mixtures *Materials and Structures* **37(267)** 202

[12] Muniz de Farias M, *et al.* 2019 Behavior of a hot-mix asphalt made with recycled concrete aggregate and crumb rubber *Canadian Journal of Civil Engineering* **46(6)** 544

[13] Rondón-Quintana H A, Ruge-Cárdenas J C, Bastidas-Martínez J, Velandia-Castelblanco M Y, Muniz De Farias M 2020 Use of thermally treated bentonite as filler in hot mix asphalt *Journal of Materials in Civil Engineering* **32(5)** 1

[14] Elseifi M A, Flintsch G W, Al-Qadi I L 2003 Quantitative effect of elastomeric modification on binder performance at intermediate and high temperatures *Journal of Materials in Civil Engineering* **15(1)** 32

[15] Dondi G, Pettinari M, Sangiorgi C, Zoorob S E 2013 Traditional and dissipated energy approaches to compare the 2pb and 4pb flexural methodologies on a warm mix asphalt *Construction and Building Materials* **47** 833

[16] Ghuzlan K A, Carpenter S H 2006 Fatigue damage analysis in asphalt concrete mixtures using the dissipated energy approach *Canadian Journal of Civil Engineering* **33(7)** 890

[17] Rao-Tangella R, Craus J, Deacon J A, Monismith C L 1990 *Summary Report on Fatigue Response of Asphalt Mixtures - Final Report to Strategic Highway Research Program, SHRP-A-312* (Washington D. C.: Transportation Research Board) p 147
[18] Carpenter S H, Shen S 2006 Dissipated energy approach to study hot-mix asphalt healing in fatigue *Transportation Research Record* **1970** 178
[19] Liu G, Jia Y, Yang T, Du H, Zhang J, Zhao Y 2017 Fatigue performance evaluation of asphalt mixtures based on energy-controlled loading mode *Construction and Building Materials* **157** 348
[20] Rondón H A, Reyes F A 2015 *Pavimentos – Materiales, Construcción y Diseño* (Bogotá: ECOE)
[21] Breysse D, De La Roche C, Domec V, Chauvin J J 2003 Influence of rest time on recovery and damage during fatigue tests on bituminous composites *Materials and Structures* **36(10)** 648
[22] Lee H J, Sias J, Kim Y R 2000 Continuum damage mechanics–based fatigue model of asphalt concrete *Journal of Materials in Civil Engineering* **12(2)** 105
[23] Kim Y R, Little D N, Burghardt R C 1991 SEM analysis on fracture and healing of sand-asphalt mixtures *Journal of Materials in Civil Engineering* **3(2)** 140
[24] Sias J, Kim Y R 2001 Laboratory evaluation of fatigue damage and healing the asphalt mixtures *Journal of Materials in Civil Engineering* **13(6)** 434
[25] Baaj H, Mikhailenko P, Almutairi H, Di Benedetto H 2018 Recovery of asphalt mixture stiffness during fatigue loading rest periods *Construction and Building Materials* **158** 591
[26] Grossegger D 2021 Fatigue damage self-healing analysis and the occurrence of an optimal self-healing time in asphalt concrete *Journal of Materials in Civil Engineering* **33(6)** 04021098
[27] Monismith C L, Secor K E, Blackmer E W 1961 Asphalt mixture behavior in repeated flexure *Journal of Association of Asphalt Paving Technology* **30** 188
[28] Taylor I F 1968 *Asphaltic Road Materials in Fatigue* (Nottingham: University of Nottingham)
[29] Pronk A C, Poot M, Jacobs M, Gelpke R 2010 Haversine fatigue testing in controlled deflection mode: is it possible? *Proceeding of the 89th Annual Meeting Transportation Research Board*, (Washington DC: Transportation Research Board) p 13
[30] Pintarelli M G, Melo J V S 201 Influence of haversine and sinusoidal wave load on estimating fatigue life of asphalt layers *Journal of Materials in Civil Engineering* **31(8)** 04019131
[31] Lundstrom R, Di Benedetto H, Isacsson U 2004 Influence of asphalt mixture stiffness on fatigue failure *Journal of Materials in Civil Engineering* **16(6)** 516
[32] Lundstrom R, Isacsson U, Ekblad J 2003 Investigation of stiffness and fatigue properties of asphalt mixture *Journal of Materials Science* **38(24)** 4941
[33] Hajj E Y, Sebaaly P E, Weitzel D 2005 Fatigue characteristics of Superpave and Hveem mixtures *Journal of Transportation Engineering* **131** 302
[34] Kocak S, Kutay M E 2021 Effect of devulcanized rubber modification on the performance grade, fatigue cracking resistance, and rutting resistance of asphalt binders *Journal of Materials in Civil Engineering* **33(9)** 04021248
[35] Júnior J L, Babadopulos L F, Soares JB 2019 Moisture-induced damage resistance, stiffness and fatigue life of asphalt mixtures with different aggregate-binder adhesion properties *Construction and Building Materials* **216** 166
[36] Clark B R, *et al.* 2018 Effects of recycled asphalt pavement on the stiffness and fatigue performance of multigrade bitumen asphalt *Journal of Materials in Civil Engineering* **30(2)** 04017278
[37] Al-Mistarehi W B, Khadaywi T S, Hussein K A 2021 Investigating the effects on creep and fatigue behavior of asphalt mixtures with recycled materials as fillers *Journal of King Saud University - Engineering Sciences* **33(5)** 355
[38] Bastidas J, Rondón-Quintana H A, Muniz de Farias M 2021 Behavior of asphalt mastics containing different materials as filler *Canadian Journal of Civil Engineering* **48** 347
[39] Wang D, Linbing W, Christian D, Zhou G 2013 Fatigue properties of asphalt materials at low in-service temperatures *Journal of Materials in Civil Engineering* **25(9)** 1220
[40] Zhou F, Mogawer W, Li H, Andriescu A 2013 Evaluation of fatigue tests for characterizing asphalt binders *Journal of Materials in Civil Engineering* **25(5)** 610
[41] Liu C, Lv S, Peng X, Zheng J, Yu M 2020 Analysis and comparison of different impacts of aging and loading frequency on fatigue characterization of asphalt concrete *Journal of Materials in Civil Engineering* **32(9)** 04020240
[42] Epps J S, Monismith C L Influence of mixture variables on the flexural fatigue properties of asphalt concrete *Proc. Association of Asphalt Paving Technology* **38** 423

[43] Bazin P, Saunier J 1967 Deformability, fatigue and healing properties of asphalt mixes *Proceeding of the Second International Conferences on the Structural Design of Asphalt Pavements* (Ann Arbor: Transportation Research Board) p 553
[44] Pell P S 1962 Fatigue characteristics of bitumen and bituminous mixes *International Conference on the Structural Design of Asphalt Pavements* (Ann Arbor: Transportation Research Board) p 43
[45] Pell P S, Taylor I F Asphaltic road materials in fatigue *Association of Asphalt Paving Technology Proceeding* **38** 371
[46] Seo Y, El-Haggan O, King M, Lee S J, Kim R 2007 Air void models for the dynamic modulus, fatigue cracking and rutting of asphalt concrete *Journal of Materials in Civil Engineering* **19(10)** 874
[47] Harvey T J, Deacon J A, Tsai B W, Monismith C L 1995 *Fatigue Performance of Asphalt Concrete Mixes and its Relationships to Asphalt Concrete Pavement Performance in California, Report RTA-65W485-2* (Berkeley: University of California) p 189
[48] Jimenez R A, Gallaway B M 1962 Behavior of asphaltic concrete diaphragms to repetitive loadings *Int. Conf. on the Structural Design of Asphalt Pavements* **203(1)** 59
[49] Doan T H 1977 Les études de fatigue des enrobés bitumineux au LCPC *Bulletin de Liaison des Laboratoires des Ponts et Chaussées* (France: LCPC) p 215
[50] Slebi-Acevedo C J, Lastra-González P, Pascual-Muñoz P, Castro-Fresno D 2019 Mechanical performance of fibers in hot mix asphalt: a review *Construction and Building Materials* **200** 756
[51] Vallerga B A, Finn F N, Hicks R G 1967 Effect of asphalt aging on the fatigue properties of asphalt concrete *Second International Conference on the Structural Design of Asphalt Pavements* (Washintong: Transportation Research Board) p 484
[52] Chen D H 2007 Field and lab investigations of prematurely cracking pavements *Journal Performance of Constructed Facilities* **21(4)** 293
[53] Sousa J, Tayebali J, Harvey P, Hendricks P, Monismith C L 1993 Sensitivity of strategic highway research program A-003A testing equipment to mix design parameters for permanent deformation and fatigue *Transportation Research Record* **1384** 69
[54] Hartman M A, Gilchrist M D, Walsh G 2001 Effect of mixture compaction on indirect tensile stiffness and fatigue *Journal of Transportation Engineering* **127(5)** 370
[55] Cooper K E, Pell P S 1974 *The Effect of Mix Variables on the Fatigue Strength of Bituminous Materials, Repor No 633* (Crowthorne: Transport and Road Research Laboratory) p 69
[56] Pell P S, Cooper K E 1975 The effect of testing and mix variables on the fatigue performance of bituminous materials *Journal Association of Asphalt Paving Technology* **44** 1
[57] Elliot R P, Thompson M R 1985 *Mechanistic Design Concepts for Conventional Flexible Pavements* (Champaign: University of Illinois) p 127
[58] Thompson M R 1987 ILLI-PAVE based full-depth asphalt concrete pavement design procedure *6th International Conference of Structural Design of Asphalt Pavements* (Ann Arbor: Transportation Research Board) p 13
[59] Finn F, Saraf C L, Kulkarni K, Nair K, Smith W, Abdullah A 1986 *Development of Pavement Structural Subsystems, National Cooperative Highway Research Program, Report 291* (Washington D. C.: Transportation Research Board) p 65
[60] The Asphalt Institute (TAI) 1982 *Research and Development of the Asphalt Institutes Thickness Design Manual (MS 1) Ninthh Edition* (Lexington: Asphalt Institute) p 204
[61] Nishizawa T, Shimeno S, Sekiguchi M 1997 Fatigue analysis of asphalt pavements with thick asphalt mixture layer *Eight Intiernational Conference on Asphalt Pavements* (Seattle: Transportation Research Board) p 969
[62] Van Dijk W, Visser W 1977 The energy approach to fatigue for pavement design *Asphalt Paving Technology* **46** 1
[63] Shell International Petroleum Company 1978 *Shell Pavement Design Manual, Asphalt Pavement and Overlays for Road Traffic* (London: Shell Institute) p 354
[64] Tayebali A A, Deacon J A, Coplantz J S, Monismith C L 1993 Modelling fatigue response of asphalt-aggregate mixtures *Journal Association of Asphalt Paving Technology* **62** 385
[65] El-Basyouny M M, Witczak M 2005 Calibration of alligator fatigue cracking model for 2002 design guide *Transportation Research Record* **1919** 76
[66] Rondón H A, Fernández W D, Noguera J A 2013 Influence of subgrade and unbound granular layers stiffness on fatigue life of hot mix asphalts - HMA *Revista TecnoLógicas* **31** 53

Influence of the formulation of a flux-cored wire on the microstructure and hardness of welded metal

E C Amaral[1], J L Jácome-Carrascal[2], A M Moreno-Uribe[3], and A Q Bracarense[3]
[1] Centro Federal de Educação Tecnológica de Minas Gerais, Timóteo, Brasil
[2] Universidad Francisco de Paula Santander, San José de Cúcuta, Colombia
[3] Universidade Federal de Minas Gerais, Belo Horizonte, Brasil

E-mail: andresmauriciomu@ufmg.br

Abstract. For this paper, the microstructure and hardness of the weld metal were investigated by conducting experiments with the flux cored arc welding process in underwater and air conditions. A rutile/oxidizing tubular wire was used, manufactured by the Robotics, Welding and Simulation Laboratory at Minas Gerais Federal University, especially for underwater wet welding. Underwater welds had a lower volumetric fraction of acicular ferrite in the weld metal compared to air welds. In the thermally affected zone, for both welds, there was a predominant formation of martensite. However, the grain size and width of the thermally affected zone of underwater welds are smaller. The hardness values shown correspond to the microstructure formed in the weld metal. On the other hand, in the region of the thermally affected zone, the hardness values were higher underwater welds, due to the smaller martensite grains presented.

1. Introduction

The underwater welding technique currently concentrates a large volume of research aimed at developing consumables, equipment, and standards of procedure for industrial applying (Uribe, *et al*, 2017) [1]. According to Moreno-Uribe, *et al.* (2020) [2], within the field of underwater wet welding (UWW), the process that has been used the most is shielded metal arc welding (SMAW), due to its characteristics such as versatility and possibility of changing the coating. However, due to its low productivity, which makes repairs more expensive, especially in deep waters, the flux cored arc welding (FCAW) process has been emerging as a promising alternative due to its work-factor, deposition rate and ease of be automated, resulting in high productivity operation, as reported by Amaral, *et al.* (2021) [3]. Also, other favorable aspect is the presence of flux, a factor that allows changing weld metal characteristics with flux composition variation (Guo, *et al,* 2016) [4].

However, even with all its potential for use in underwater wet welding, the application of tubular wires has still been limited, as mentioned by Castellanos-González, *et al.* (2021) [5]. Few studies address the use of this process in underwater wet welding, due to the difficulty in developing consumables and equipment (welding torches) (Assunção and Bracarense, 2017) [6], in understanding and/or combining the characteristics that allow associating its high productivity with adequate mechanical properties of the joint, according to Mendonça and Bracarense (2019) [7].

Thereby, given the promising results obtained by the tubular wires reported by Amaral, *et al.* (2021) [3], regarding the stability of the electric arc and the reduction of diffusible hydrogen in the weld metal, and considering the growing demand for processes that ensure high productivity in wet underwater welding. This work has as objective the analyses of an experimental tubular wire, as for application in underwater wet welding; given the reasons mentioned, the present work aims to evaluate the influence of a self-protected tubular wire with a rutile-oxidizing base on the microstructure and hardness of the weld metal.

2. Methodology and materials

The experiments conducted in this work were carried out with a 1.6 mm rutile/oxide self-protected tubular wire developed by the Robotic, Welding and Simulation Lab (LRSS), Brazil, especially for UWW. The flux composition of manufactured wires is summarized below (see Table 1).

Table 1. Formulation of experimental flux-cored wires (wt %).

Flow	TiO_2	SiO_2	Fe_2O_3	FeSiMn	CaO	FeTi	Iron powder
F0	35	14	14	12	3	1	21

All experiments were carried out at a depth of 0.5 m in a flat position in a tank; during welding, the torch was kept stationary and a welding table with the test specimen moved at a speed of 4 mm.s-1. In this study, an ESAB welding source, model AristoMIG 5001i, in constant voltage mode was used to keep the arc voltage constant during the welding process. The wire feed speed, welding speed and the contact-tip to work distance were fixed. The welding current was automatically changed by the welding power source in order to maintain the pre-set arc voltage and keep the wire feed speed and the fusion rate in the same proportion, as mentioned by Castellanos-González, *et al.* (2021) [8]. Six weld beads were made for each welding condition (air and underwater condition), with the same welding parameters, resulting 12 weld beads. Table 2 presents the welding parameters used.

Table 2. Welding parameters.

Voltage	28 V
Wire feed speed	4.5 m/min
Speed of welding	250 mm/min
CTWD*	30 mm
Polarity	DCEP**

* CTWD: contact-tip to work distance.
** DCEP: Direct current electrode positive.

Based on the work developed by Moreno-Uribe, *et al.* (2020) [9] A data acquisition system was used with 5 KHz acquisition frequency, to monitor the welding current, arc voltage and wire feed speed. Figure 1 schematically presents an overview of the entire system necessary for the acquisition of welding data.

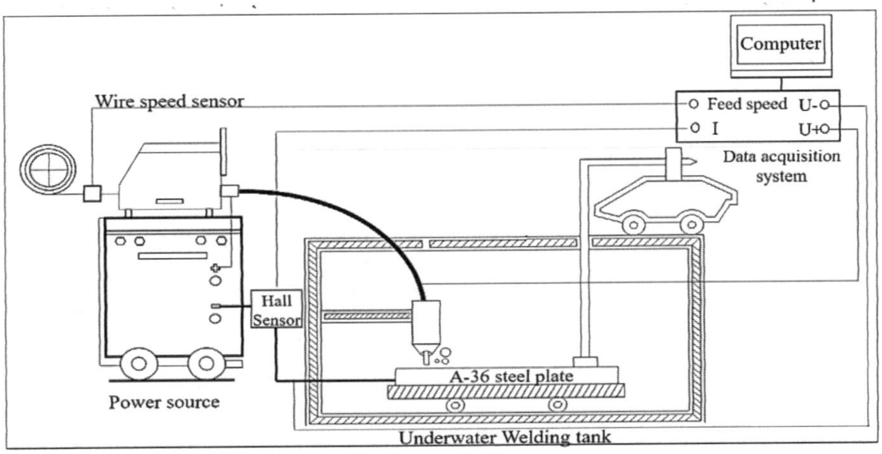

Figure 1. Diagram of welding jig (unscaled).

3. Results and discussions

The results of the project show two types of analysis, structural analysis, and hardness analysis.

3.1. Microstructural analysis

To characterize the microstructure, the volumetric fraction of acicular ferrite formed in the weld metal and the other constituents formed was calculated. The methodology used in the analysis was proposed by the International Institute of Welding (IIW) doc. IX-1533-88 [10]. Samples were taken from three different regions of beads deposited on an ASTM A36 steel plate [11]. After the metallographic preparation, the images were obtained with the aid of a CCD camera model Hyper-HAD from the manufacturer Sony coupled to an optical microscope model BX60MF5 from the manufacturer Olympus. For image analysis, Image J software was used. Figure 2 and Figure 3 present the typical microstructures found in the welds of the two welding conditions, in air and UWW, respectively. The microstructures of the top, center, and base of the fused zone and the HAZ of each weld bead are shown.

Figure 2(a) to Figure 2(c) and Figure 3(a) to Figure 3(c), show the formation of different microconstituents in the fused zones of the welds in both welding conditions. The volume fraction of each constituent is shown in Table 3. In the molten zone (see Figure 2(a), Figure 2(b), and Figure 2(c)), it is possible to observe a greater uniformity in the microstructure of the air weld and a predominance of acicular ferrite. In UWW, Figure 3(a), Figure 3(b), Figure 3(c), there is greater heterogeneity in the microstructure and without the massive predominance of a particular constituent. This condition may be linked to the greater number of thermal processes and high cooling rates related to UWW, which influences the reduction of the austenitic grain size.

This result is in line with those presented by Zhang, *et al.* (2016) [12], where underwaters welds presented a lower proportion of acicular ferrite and pro-eutectoid ferrite compared to air welds, with all welds performed with the same welding parameters and by the FCAW process. In both welding conditions (see Figure 2(d) and Figure 3(d)), the heat affected zone showed a predominance of martensite (M) formation. However, the grain size and ZTA width of underwater welds are smaller (see Figure 3(d)).

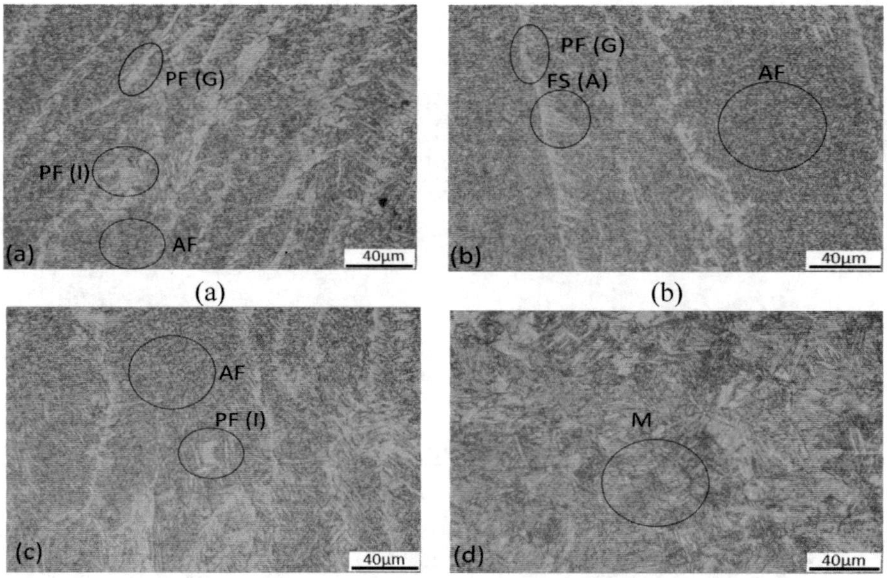

(a) (b) (c) (d)

Figure 2. Fused zone: (a) top, (b) center and (c) base. HAZ: (d). Composed of acicular ferrite (AF), primary grain boundary ferrite (PF(G)), polygonal intragranular primary ferrite (PF(I)) and aligned second-phase ferrite (FS(A)).

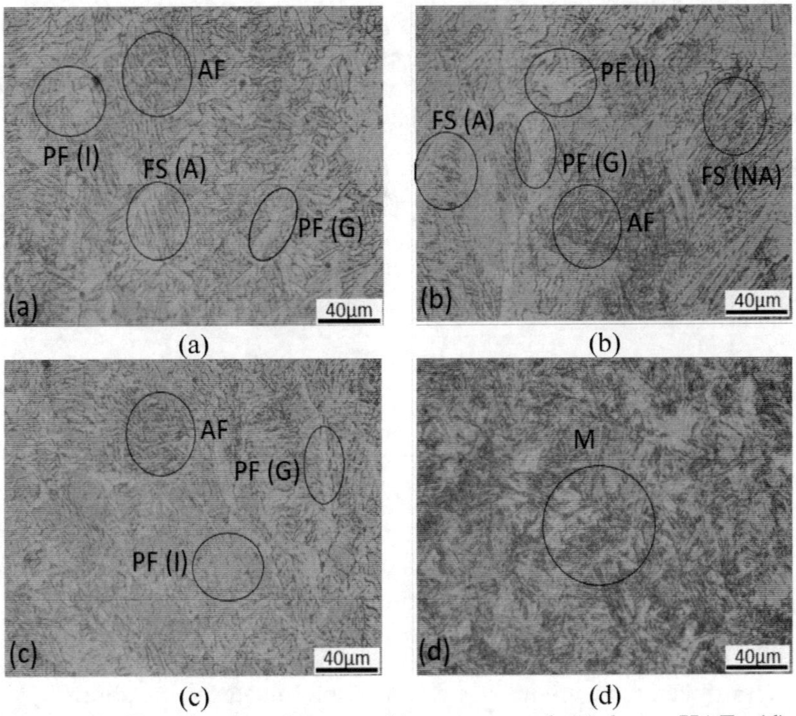

Figure 3. Fused zone: (a) top, (b) center, and (c) base. HAZ: (d). Composed of acircular ferrite (AF), grain boundary primary ferrite (PF(G)), polygonal intragranular primary ferrite (PF(I)), aligned second-phase ferrite (FS(A)) and non-aligned second-phase ferrite (FS(NA)).

Table 3. Volume fraction of the constituents present in the ZF and HAZ.

Analysis region	Microconstituent	Weld bead	
		Air	UWW
ZF	AF	46.0 ± 2.0	16.5 ± 0.5
	PF (G)	20.5 ± 1.5	22.5 ± 0.5
	PF (I)	17.0 ± 1.0	14 ± 0.0
	FS (A)	13.0 ± 1.0	32 ± 2.0
	FS (NA)	3.5 ± 0.5	15 ± 1.0
HAZ	M	83.5 ± 1.5	89.5 ± 0.5
	B	16.5 ± 1.5	10.5 ± 0.5

Acircular ferrite (AF), grain boundary primary ferrite (PF(G)), polygonal intragranular primary ferrite (PF(I)), aligned second-phase ferrite (FS(A)) and non-aligned second-phase ferrite (FS(NA)), martensite (M) and bainite (B).

3.2. Hardness analysis

Figure 4 presents the hardness profile, for welding in air and underwater conditions, from the top of the bead to the base metal. It can be observed that the microhardness values obtained for the three regions correspond with the presented microstructures. Air welds had lower hardness in the ZF region, as they presented higher volumetric fractions of the toughest constituent, acicular ferrite, than UWW welds.

On the other hand, in the HAZ region, the two welding conditions have the same constituent, martensite, but in the UWW welds, due to the higher cooling rate, the grains had a smaller size, ensuring greater hardness. The MB region had similar hardness for all conditions due to the use of the same material for all welds.

The results presented for microhardness in the ZF and HAZ of the weld beads in the two welding conditions (air and UWW) are superior to those presented by Zhang et al (2016) [12] also in the two welding conditions. In comparison of microhardness in ZF and HAZ, only for the UWW welding condition, the results are similar to those presented by Gao, *et al.* (2016) [13]

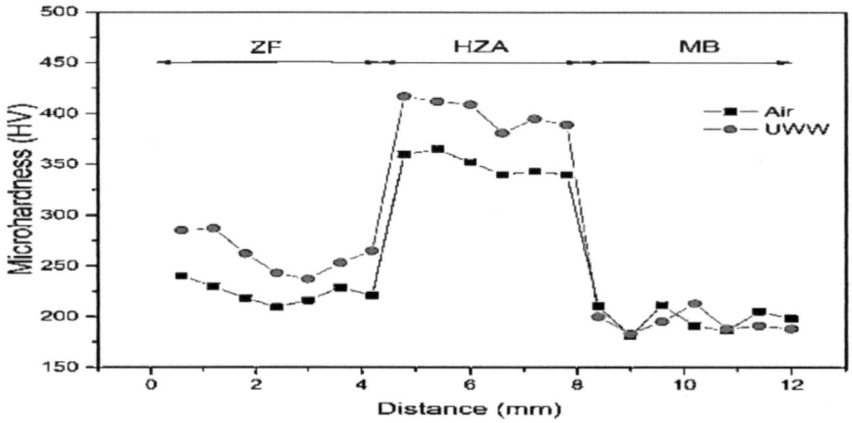

Figure 4. Vickers microhardness profile from HAZ, ZF and MB.

4. Conclusions

In this work, an experimental tubular wire was designed to evaluate the metallurgical characteristics of the weld metal and the heat-affected zone, in air and underwater wet welding. UWW welds had less AF in the molten zone compared to air welds; the HAZ, for both conditions, showed a predominance of martensite, and in the UWW condition, the grain size was apparently smaller.

The results of the Vickers microhardness are in line with the microstructure formed, where due to the smaller volumetric fraction of AF in the molten zone of the UWW welds, the hardness presented a lower value. On the other hand, in the HAZ, the hardness presented high values for both conditions in relation to the molten zone and a slight increase in the hardness of the HAZ in the UWW condition in relation to the air condition.

Acknowledgments

The authors would like to acknowledge the Laboratório de Robótica, Soldagem e Simulação (LRSS), Brazil, for the availability of resources for the accomplishment of this work.

References

[1] Moreno Uribe A M, Queiroz Bracarense A, Pereira Pessoa E C, Rocha dos Santos V 2017 Influência da polaridade sobre a estabilidade do processo de soldagem subaquática molhada com eletrodo revestido *Soldagem & Inspeção* **22(4)** 429

[2] Moreno Uribe A M, Bracarense A Q, Pessoa E C 2020 The effect of polarity and hydrostatic pressure on operational characteristics of rutile electrode in underwater welding *Materials* **13** 5001

[3] Amaral E C, Moreno Uribe A M, Bracarense A Q 2021 Effects of PTFE on operational characteristics and diffusible H and O contents of weld metal in underwater wet welding *Journal of Manufacturing Processes* **61** 270

[4] Guo N, Xu C, Du Y, Wang M, Feng J, Deng Z, Tang D 2016 Effect of boric acid concentration on the arc stability in underwater wet welding *Journal of Materials Processing Technology* **229** 244

[5] Castellanos González O, Prada Botia G, Jácome-Carrascal J, Moreno Uribe A, Rodríguez Arias A 2021 manufactura de alambre tubular autoprotegido y evaluación de condiciones operacionales en soldadura submarina *Investigación e Innovación en Ingenierías* **9(1)** 167

[6] Assunção M T, Bracarense A Q 2017 Evaluation of the effect of the water in the contact tip on arc stability and weld bead geometry in underwater wet FCAW *Soldagem & Inspeção* **22** 401

Journal of Physics: Conference Series **2118** (2021) 012010 doi:10.1088/1742-6596/2118/1/012010

[7] Mendonça C M, Bracarense A Q 2019 Investigação do fenômeno das bolhas em soldagem subaquática molhada com arame tubular autoprotegido *Soldagem & Inspeção* **24** e2416:1

[8] Castellanos González O M, Moreno Uribe A M, Ramón Ramón S A, Jácome J L 2021 Evaluación de la transferencia metálica y estabilidad del proceso GMAW *Revista UIS Ingenierías* **20(3)** 47

[9] Moreno Uribe A M, Romanelli Sales C G A, Ribeiro Menezes P H, Queiroz Bracarense A 2020 Study of the transition current in GMAW (MIG/MAG) *Ingeniare. Revista Chilena de Ingeniería*, **28(3)** 528

[10] Abson D J 1988 Guide to the light microscope examination of ferritic steel weld metals *IIW Doc. No. IX-1533-88, IXJ-123-87 Revision 2* (Miami: American Welding Society)

[11] American Society for Testing and Materials (ASTM) 2019 *Standard Specification for Carbon Structural Steel, ASTM A36/A36M-19* (United States of America: American Society for Testing and Materials)

[12] Zhang Y, Jia C, Zhao B, Hu J, Wu C 2016 Heat input and metal transfer influences on the weld geometry and microstructure during underwater wet FCAW *Journal of Materials Processing Technology* **238** 373

[13] Gao W, Wanga D, Chenga F, Dia X, Denga C, Xu W 2016 Microstructural and mechanical performance of underwater wetwelded S355 steel *Journal of Materials Processing Technology* **238** 333

Physical and mechanical behavior of fine soil according to the content of multispecies diatoms

D A Zuluaga[1], D Sabogal[1], C A Buenaventura[2], and C J Slebi[3]
[1] Escuela de Ingenieros Militares Ejército de Colombia, Bogotá, Colombia
[2] Concrelab SAS, Bogotá, Colombia
[3] Universidad de Cantabria, Santander, España

E-mail: daniel.zuluaga@esing.edu.co

Abstract. The physical response and geotechnical properties of diatomaceous soils are not fully understood, data are sparse, and do not account for the effects of single and multispecies frustule content, origin, type, and variability. The main physical problem lies in the irregular response of diatomaceous soils due to micro and nano scale causes and its unexpected effects on the macro scale. This research compared the characteristics of a multispecies diatomaceous soil sample (North American origin) with other diatomaceous single-species soils. Six artificial soil mixtures were prepared, dosed by weight, in order to determine the influence of the content of frustules. The results show that the liquid limit of the samples is lower than that of the monospecies samples for any content of frustules. The pore areas of the monospecies samples are found to be 4 to 7 times larger than those of the North American soil. Void ratios and compressibility ranges are higher as the diatom content increases. The internal friction angle of diatomaceous soils varies in a non-linear tendency with respect to fossil content. For the studied soil at 100% fossil concentration, the internal friction angle reached 38.32°, a magnitude that is lower than the values reported for most of the monospecies contrast samples.

1. Introduction

Diatoms are microscopic unicellular photosynthetic algae that grow in terrestrial and aquatic environments of fresh or saltwater [1], rich in dissolved silica, whose existence depends on the volcanic activity of the area [2-4]; After their death and organic decomposition, the frustules are deposited on the bottom of oceans or lakes, forming diatomaceous soils [1,5,6]. These soils have been found in different parts of the world and have been scarcely studied in physics and geotechnics; of particular interest is the cell wall of diatoms (frustules), which is formed of silica [7] and gives rise to nanometric structures [8]. Species evolution has created distinctive patterns of perforations in the frustules, which are used for visual classification [9].

Physics and soil mechanics dictate that, the greater the irregularity in the surface of a particle "A" that forms a continuous medium, the greater the resistance to differential displacement with particle "B", therefore, the greater the mechanical response of the system. This is easy to understand and demonstrate at larger scales (> 0.075 mm), but it is not so much in nano and micro sizes, where according to geotechnics, everything is classified as soft soil and far from any high resistance parameter. The physical problem for diatomaceous soils lies in the fact that phenomena like secondary consolidation or interlocking of particles on a micro scale (interlocking) are not fully understood, which are presumed to be related with the quantity, shape (according to the species-mono or multi) and the state of conservation of the frustules.

Some of the properties of diatomaceous soils (DS) are high porosity, high absorption [10-12], high initial void ratio, and low density [4,13]. High liquid limit values have been identified in diatomaceous soils from Mexico and Japan [2,14]. Values of void content have been reported between 80% and 90% [15,16], and "specific surface area" magnitudes have been registered between 40 and 350 m^2/g [2,17]. However, in none of the documentary sources, the physical and mechanical behavior is analyzed according to the type (one = mono, or several = multi), concentration, or species. Additionally, [18] highlights that the presence of diatoms (microscopic skeleton of algae and dead plankton) in soils and sediments leads to the measurement of mechanical characteristics in magnitudes that contradict the behaviors defined in classical geotechnics.

Soils with the presence of diatoms can retain high water contents inside their frustules [4,19] and simultaneously record high values of friction angle and high shear strength [4,19,20]. According to Cheng, *et al.* [21] the presence of diatoms leads to an increase in Atterberg limits and compressibility, and it allows high pore pressures and low effective confining pressures during shear tests. Cheng mentions that the deformation mechanisms of microstructural elements may explain the unusual behavior of this type of soils.

The main target of this research was to compare the physical and mechanical characteristics between a multispecies diatomaceous soil (North American origin) and other diatomaceous single-specie soils. Six artificial soil mixtures were prepared, dosed by weight, to determine the influence of the content of frustules. The main evaluated parameters were particle size distribution, specific gravity (SG), pore frustule areas, consistency, compressibility, and shear resistance.

2. Materials and methods

Artificial soil samples were prepared by the weight-controlled combination of a fine matrix of Kaolin type and monospecies diatomaceous soils (MNDS), to know the influence of the content, type, and mode of diatoms (monospecies or multispecies) on the shear strength and volumetric variation. The results of the MNDS were compared with data from previous studies, concerning samples of diverse origins, mostly of monospecies conditions. Figure 1 shows the particle size distribution of the MNDS, and the kaolin used; the curves of other single and multispecies diatomaceous soils are also contrasted.

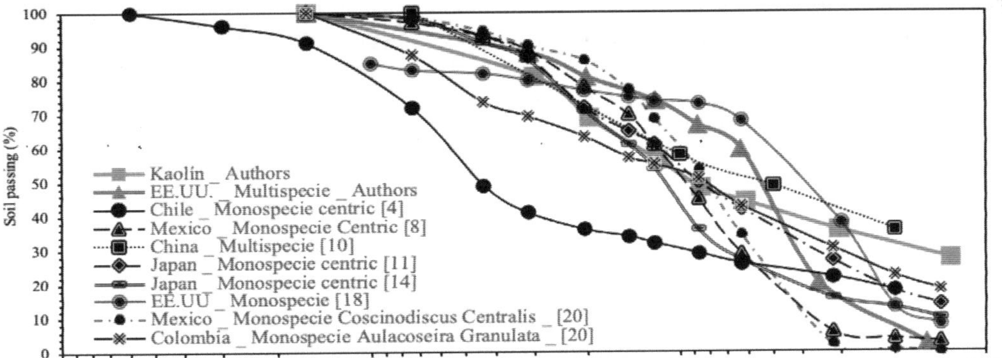

Figure 1. Contrast of particle size distribution of diatomaceous soils and kaolin.

The MNDS reports particle sizes corresponding to a silt (>2μm), its d50 is 0.0055 mm; Kaolin presents a uniform curve in the fine particle range (< 0.075 mm), while the MNDS records a particle concentration of 39.16% between 3.6 μm and 6.3 μm. The d50 of Kaolin is 0.0095 mm. (d50: particle size corresponding to 50% that passes, represents the average grain size). No differential pattern is identified in the multispecies sample with respect to the other monospecies references.

Figure 2 presents the variation of the SG of the MNDS samples regarding the diatomaceous content. SG values of MNDS are slightly lower than the Mexican sample (monospecie coscinodiscus centralis). It is evident the distance with Colombian sample (monospecies aulacoseira granulata). Table 1 shows the SG values of other diatomaceous soils.

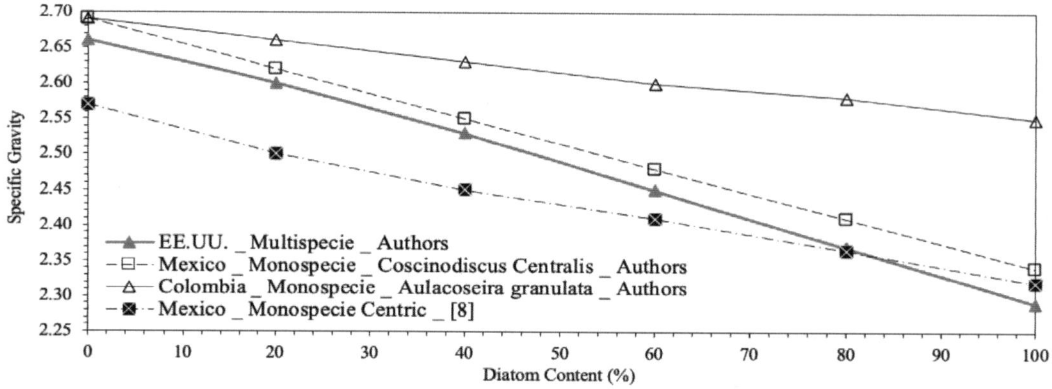

Figure 2. SG according to diatom content.

Table 1. SG reference values in diatomaceous soils.

Diatomaceous soil origin	Type	SG	
Mexico	Not defined	2.72 – 2.80	[3]
Chile	Centric monospecies	2.63	[4], [5], [19]
Indonesia	Not defined	1.87 – 2.00	[7]
Japan	Not defined	2.40 – 2.70	[8]
Mexico	Centric monospecies	2.32	[8]
Mexico	Not defined	2.60	[12]
USA	Monospecies	2.08	[18]

The Moisture content applied during the sample preparation process was the ratio of 1.5 times the liquid limit, which varied according to the proportion of solids, see Table 2; from the extracted samples (radius: 80 mm and height: 200 mm), specimens were cut and subsequently used for the execution of direct shear and consolidation tests. A pre-consolidation vertical stress of 30 kPa was applied. The shear test was carried out in drained consolidated conditions, at a speed of 0.05 mm/min; for the one-dimensional consolidation test, the vertical stress range was between 7 kPa and 1,600 kPa. The MNDS sample was imaged with scanning electron microscopy (SEM) [23] in order to recognize the surface texture [7], morphology (size and shape) [7], and identify the species of the microfossils.

Table 2. Dosage and characterization of artificial soil (Kaolin + MNDS).

Diatom dosage (weight/weight)	Diatoms content (g)	Kaolin content (g)	Total solids (g)	Water content (ml)	Specific gravity	Liquid limit (%)	Plastic limit (%)	Plasticity index	Activity
100% D	590	0	590	785	2.29	89	61	28	3.9
80% D	566	142	708	828	2.37	78	58	20	1.7
60% D	496	330	826	818	2.45	66	48	18	1.0
40% D	378	566	944	810	2.53	57	39	18	0.8
20% D	212	850	1062	744	2.60	47	30	16	0.6
0% D	0	1180	1180	600	2.66	34	22	12	0.4

3. Results and analysis

The results of microscopical characterization and physic-mechanical tests applied to the contrast soils (multi and mono specie samples) are presented below.

Journal of Physics: Conference Series **2118** (2021) 012011 doi:10.1088/1742-6596/2118/1/012011

3.1. Scanning electron microscopy

SEM was used to visualize the surfaces and structures of diatomaceous fossils in nano and micro scales. Four species were identified in MNDS (see Table 3). Two examples of identified species are presented in Figure 3. For each specie, the surface irregularities were identified (see Figure 4). Pores of each specie were characterized in density (number of pores/μm^2) and pore area (μm^2). Pores constitute potential water collectors (see Figure 5).

As for the pore area, the MNDS results were compared with two monospecies samples. The pore areas of the monospecies samples were found to be 4 to 7 times larger than the pore areas of any multispecies sample. On the other hand, the monospecies frustules report lower pore density than any of the species identified in the multispecies sample (see Table 3).

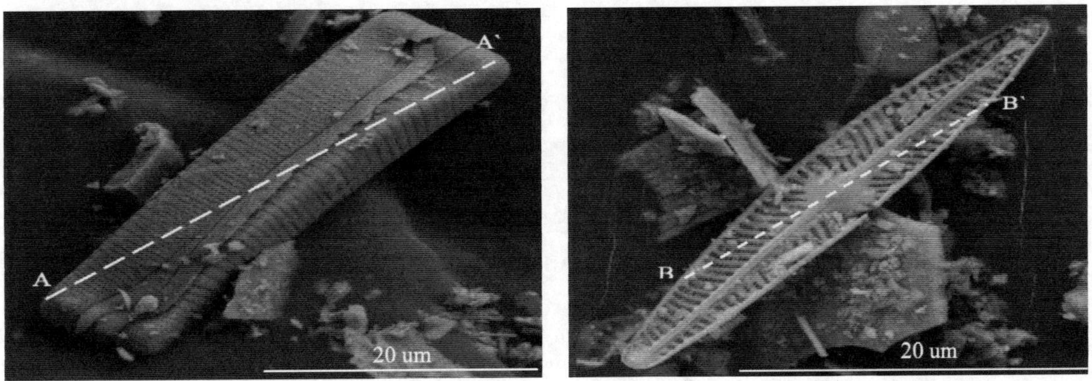

(a) (b)

Figure 3. Species identification within the diatomaceous soil sample of North American origin; (a) orizaformis holarctica, (b) navicula cryptocephala.

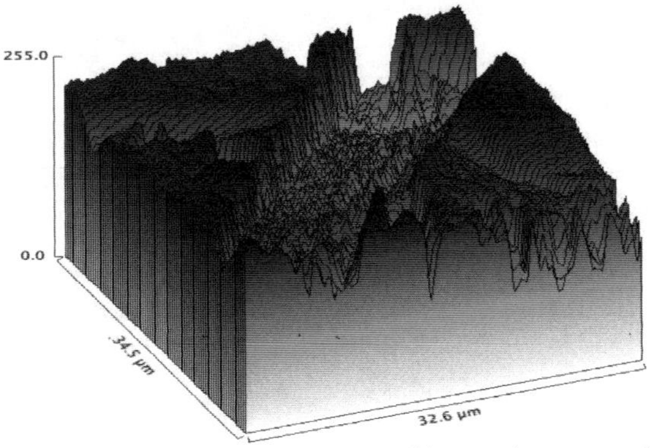

Figure 4. Example of Profile and area of Irregularities on the surfaces of the frustule orizaformis holarctica specie.

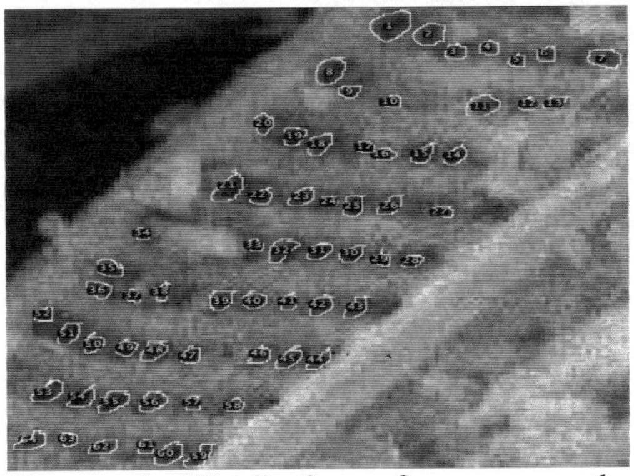

Figure 5. Distribution of pores on the surfaces of the frustules navicula cryptocephala specie.

Table 3. Properties North American multispecies diatoms.

Species	Type	Average geometric characteristics (μm)			Density
		Length	Width	Height	(pores/μm^2)
Orizaformis holarctica	Multiespecies	41.68	9.11	2.69	5.0
Navicula cryptocephala	Multiespecies	37.40	5.35	0.31	5.0
Achnanthidium minutissimum	Multiespecies	12.56	2.03	1.08	8.0
Nitzschia soratensis	Multiespecies	17.26	4.79	1.04	4.0
Aulocoseira granulata	Monoespecies	17.66	12.13	12.13	1.5
Cosconodiscus centralis	Monoespecies	15.62	15.62	5.02	3.0

3.2. Consistency limits

From Figure 6, it can be concluded that the classification using the plasticity chart depends on the diatom concentration; it is observed that the increase of the MNDS, implies the increase of the plasticity index (PI). In some monospecies samples (Mexican and Colombian [20]), the increase of PI regarding liquid limit (LL) is observed parallel to line A in the plasticity chart. On the contrary, the almost perpendicular behavior is evidenced in the monospecies samples from Japan [14] and Mexico [8]. The MNDS sample, although not completely parallel to line A, does preserve a proportional behavior between PI and LL. Usually, a higher PI value in a soil is associated with a lower shear stress resistance. However, this phenomenon is the opposite in the diatomaceous soils analyzed, which, while increasing their PI, also registered a higher friction angle, as the fossil content increased.

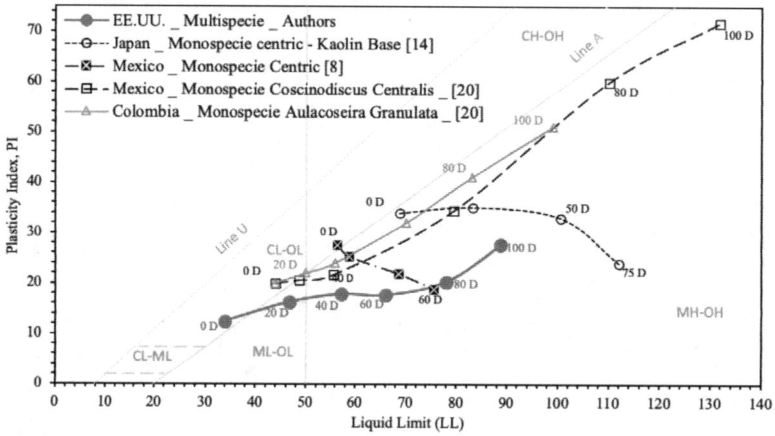

Figure 6. Classification of different diatomaceous soil samples from the plasticity chart.

3.3. Compressibility analysis

Figure 7 presents variation of the void ratio is recorded between 0.95 and 3.7; a particular case is found in the Japanese soil (monospecies centric condition) [14], which registers values as high as 6.75. For Colombian (lake origin) [13] and Chilean (centric monospecies) [4] species, the fringes were projected for different stress levels, which were found to fit within the typical compressibility ranges for DS. For the MNDS a maximum void ratio of 1.9 is reported (100% concentration).

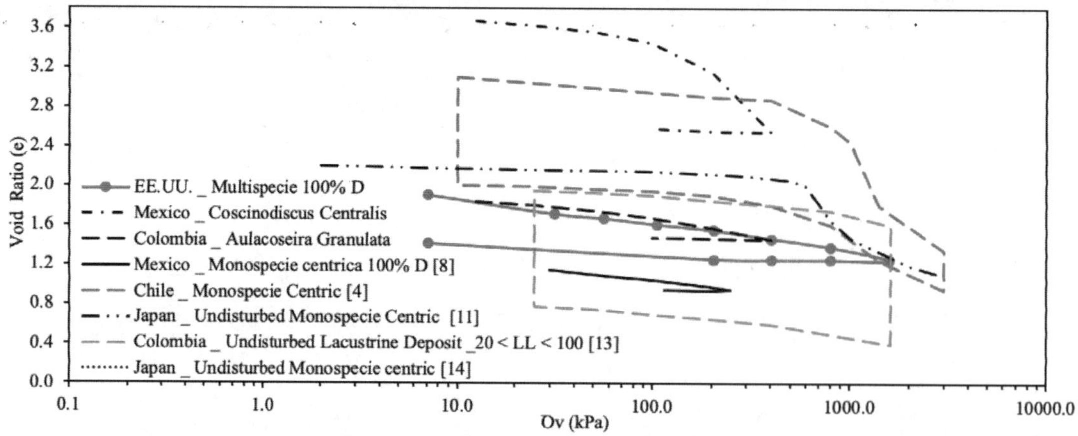

Figure 7. One-dimensional consolidation curves for different diatomaceous soil samples.

Journal of Physics: Conference Series **2118** (2021) 012011 doi:10.1088/1742-6596/2118/1/012011

3.4. Shear resistance using direct shear test

The internal friction angle (IFA) of diatomaceous soils varies in a non-linear way respect to the fossil content (see Figure 8). For all cases, the higher the content of fossils, the higher the IFA. For MNDS (100% concentration), the IFA reached values of 38.32°, which, although a considerable record for a fine soil, is lower than the IFA reported in most of the monospecies contrast samples (Mexican origin 50.8° and 51.5° for concentrations of 100% and 60%, respectively).

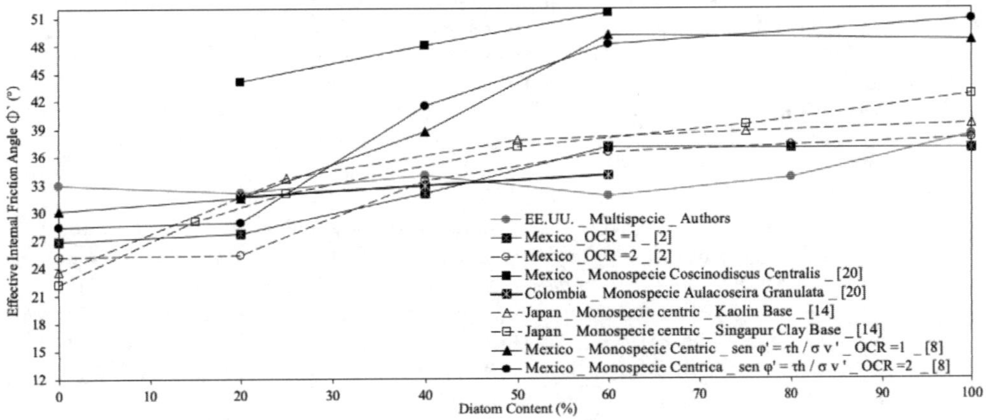

Figure 8. Variation of friction angle respect to diatom content.

4. Conclusions

The presence of microfossils in soils causes an increase in moisture; there is a proportional relationship between liquid limit and plastic limit with diatom content. The liquid limit of the MNDS is found to be lower than the monospecies samples for any frustules content. As the soil presents higher MNDS dosages, a greater volumetric change is observed during consolidation. According to physics it could be understood because the structures of the fossils provide higher void content, not only because of the spaces created by the totality of the frustule, but also because of the pores present in them. The resistance to shear stress represented by internal friction angle is proportional to the content of microfossils in the soil, reaching values of up to 38.32°. This is explained by factors such as geometry, skeletal interlocking, and the rough surface. However, the MNDS results are lower than the values associated with monospecies samples. From above, it is concluded that the physical properties of frustules influence in the mechanical response on a macro scale, and that variability of the species is a determining factor in the response to imposed stresses.

References

[1] Flower R 2013 Diatomites: their formation, distribution, and uses *Earth Systems and Environmental Sciences* **2** 501

[2] Díaz J 2011 Comportamiento monotónico de suelos diatomáceos *Rev. Int. de Desastres Naturales, Accidentes e Infraestructura Civil* **12(1)** 27

[3] Mosquera A, León L 2018 *Comparación del Comportamiento Físico-Mecánico de las Arcillas Bentoníticas Provenientes de Bugalagrande, Valle del Cauca, Colombia con Adición de Diatomeas de la Ciudad de México* (Bogotá: Universidad Católica de Colombia)

[4] Arenaldi G, Ovalle C, Barrios A 2019 Compressibility and creep of a diatomaceous soil *Engineering Geology* **258** 1

[5] Arenaldi G, Ovalle C 2019 Compresibilidad y propiedades dinámicas de suelos diatomáceos de Mejillones *Obras y Proyectos* **25** 6

[6] Mejía J, Mejía De Gutiérrez R, Montes C 2016 Rice husk ash and spent diatomaceous earth as a source of silica to fabricate a geopolymeric binary binder *Journal of Cleaner Production* **118** 133

[7] Hasan M, *et al.* 2020 Characteristic of calcined diatomaceous earth from Aceh Besar District - Indonesia as cementitious binder *IOP Conf. Ser.: Mater. Sci. Eng.* **933** 012008:1

[8] López J 2009 *Comportamiento Cíclico de Suelos Diatomáceos* (Ciudad de México: Universidad Nacional Autónoma de México)
[9] Lozano Y 2010 Listado de diatomeas (bacillariophyta) registradas para el mar caribe colombiano *Bol. Invest. Mar. Cost.* **39(1)** 83
[10] Zhang Y, Guo C, Yao X, Qu Y, Zhou N 2013 Engineering geological characterization of clayey diatomaceous earth deposits encountered in highway projects in the Tengchong region, Yunnan, China *Engineering Geology* **167** 95
[11] Tanaka H, Locat J 1999 A microstructural investigation of Osaka Bay clay: the impact of microfossils on its mechanical behaviour *Canadian Geotechnical Journal* **36(3)** 493
[12] Díaz J, González R 2013 Influence of diatom microfossils on soil compressibility *Proceedings of the 18th Int. Conf. Soil Mech. Geotech. Eng. Challenges Innov. Geotech.* (Paris: ICSMGE) p 325
[13] Caicedo B, Mendoza C, López F, Lizcano A 2018 Behavior of diatomaceous soil in lacustrine deposits of Bogotá, Colombia *Journal of Rock Mechanics and Geotechnical Engineering* **10(2)** 367
[14] Shiwakoti D, Tanaka H, Tanaka M, Locat J 2002 Influences of diatom microfossils on engineering properties of soils *Soils and Foundations* **42(3)** 1
[15] Aksakal E, Angin I, Oztas T 2012 Effects of diatomite on soil physical properties *Catena* **88(1)** 1
[16] Khraisheh M, Al-degs Y, Mcminn W 2003 Remediation of wastewater containing heavy metals using raw and modified diatomite *Chemical Engineering Journal* **99(2)** 177
[17] Wu J, Yang Y, Lin J 2005 Advanced tertiary treatment of municipal wastewater using raw and modified diatomite *Journal of Hazardous Materials* **127(1)** 196
[18] Palomino A, Kim S, Summitt A, Fratta D 2011 Impact of diatoms on fabric and chemical stability of diatom-kaolin mixtures *Applied Clay Science* **51(3)** 287
[19] Ovalle C, Arenaldi G 2020 Mechanical behaviour of undisturbed diatomaceous soil *Marine Georesources and Geotechnology* **39(5)** 623
[20] Caicedo B, Zuluaga D, Slebi C 2019 Effects of micro-features of fossil diatom on the macroscopic behaviour of soils *Geotechnique Letters* **9(4)** 322
[21] Cheng X, Ngan-Tillard D, Den Haan E 2007 The causes of the high friction angle of dutch organic soils *Engineering Geology* **93(1)** 31

Implications of physical and chemical changes in volcanic ash soils related to road construction

L R Vásquez-Varela[1], and C Mendoza[1]
[1] Departamento de Ingeniería Civil, Universidad Nacional de Colombia, Manizales, Colombia

E-mail: lrvasquezv@unal.edu.co, cmendozab@unal.edu.co

Abstract. Soils derived from weathered pyroclastic material in a humid tropical environment, known as volcanic ash soils, cover a low portion of the Earth's surface. However, its unique characteristics, such as high porosity, low unit weight, high water content, and susceptibility to change in properties by drying and remolding, justify its study due to the impact of these characteristics in the construction of embankments and road subgrades, particularly on the compaction characteristics. The urban settlements located in the tropical zone, close to areas of volcanic activity, develop their road infrastructure on this type of soil. Under laboratory conditions, these soils' chemical and physical changes, particularly on the particle arrangement or fabric, create a gap between their expected and actual behavior at the construction site. This article presents some problems in road construction related to these unique features and how to deal with them on the construction site, applying conventional classification and compaction tests in the lab and construction site. The particularities of the soils in these conditions are explained using unconventional tests to assess the microscopic effects of compaction on the soil fabric.

1. Introduction

Volcanic ash soils are less than 1% of the Earth's continental surface [1]. However, the effect of its genesis on its geotechnical behavior [2] and the subsequent impact on infrastructure justifies the study of these soils, particularly in those aspects in which they differ from the expected behavior of the sedimentary soils covered in traditional soil mechanics.

This work presents some problems in road construction on and with soils weathered from volcanic ash in a hot-humid tropical environment. Several researchers studied the characteristics of this type of soil. Numerous works emphasize the difficulties of classifying soils due to their change in properties because of the standard laboratory handling processes [3-8]. Other authors investigated the mechanical properties of these soils to establish their stability in slopes and foundations [6,9-11]. Also, the seismic response of these soils has been investigated [12,13].

The authors review some experiences on road construction issues related to the classification of volcanic ash soils and the obtaining of relationships between dry unit weight and moisture in road construction quality assurance and control. A single physical parameter, dry unit weight, is key to quality assurance and construction control; however, due to chemical and physical changes of volcanic ash soil under drying, this value must be assessed under controlled conditions to avoid unrepresentative results.

2. Background

As previously stated, residual soil deposits from weathered volcanic ash are a small fraction of the world's soils [1,14]. Although less studied than sedimentary soils, volcanic ash soils present problems such as cracking, high water retention capacities, workability problems due to excessive moisture, and very soft consistencies that justify its study [1,2,10,15-17]. In addition, the properties of these soils vary depending on the wetting or drying path [2,18-19]. These problems imply challenges in constructing embankments and road subgrades on these soils, particularly with the compaction characteristics used in earthworks quality assurance and quality control.

Figure 1(a) shows that soil deposits weathered from volcanic ash occupy approximately 11.6% of Colombia's territory on the three branches (Cordilleras) of the Andes [1]. This region comprises a vast majority of the economic activities and infrastructure development.

Figure 1(b) and Figure 1(c) show the formation and transformation of andesitic volcanic ash into soil. First, smaller than 2 mm ash particles are expelled by the volcanic eruption and transported by the wind. In Colombia, volcanic ash soil originated from the Ruiz, Tolima volcano complex eruptions in the last 20 thousand years of the Holocene and Pleistocene epochs of the Quaternary [8]. Once deposited, the volcanic ash suffers a transformation or diagenesis phase into volcanic ash soil. The mineralogy of these soils contains unusual clay minerals and amorphous materials, such as allophane, which are products of very rapid cooling that inhibit the development of a crystalline structure.

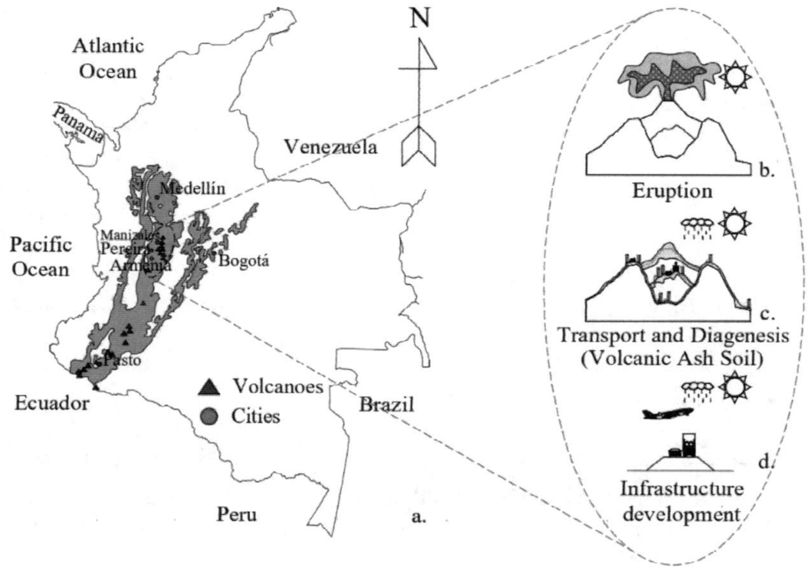

Figure 1. Volcanic ash soils in Colombia: (a) geographical distribution, (b) to (d) formation process; adapted from [1,20].

Allophane consists of hollow spheres that measure approximately 5 nm in diameter with 3.8 nm internal diameters and holes of about 0.3 nm [21]. These allophane spheres are mixed with imogolite, which consists of fine thread-like fibers. Weathering of allophane produces halloysite, which also has unusual properties. Additional weathering transforms halloysite into kaolinite [2]. These minerals are obtained by physical and chemical processes, including dissolution, leaching, and precipitation of the initial volcanic ash. In general, the ash minerals are light ($G_s < 2.8$), and the soils have high void ratios, porosities, and liquid limits. Also, a brittle structure is formed that changes in the compaction process leading to a change in behavior [2]. This behavior change is different depending on the wetting or drying path.

Figure 2 was taken by authors' request at the Instituto de Investigaciones en Estratigrafía (IIES) at Universidad de Caldas, Colombia, using scanning electron microscopy (SEM) for an undisturbed sample, and a compacted sample of volcanic ash soil from Manizales, Colombia, obtained at geographical coordinates 5°03'25.5" N / 75°28'49.9" W and three meters depth.

Regarding mechanical properties of volcanic ass soils, results showed by Hernández, et al. [2], Herrera, et al. [8], Arango [9], Forero-Dueñas, et al. [11], Yamin, et al. [12], Wesley [22], and Moroto [23] report peak effective friction angle of residual soils derived from volcanic ash in a range from 25 to 43 degrees, which increases with the compaction process [2]. Herrera, et al. [8] and Wesley [22] report residual friction angles with lower values. Lizcano, et al. [1] made vane tests on volcanic ash soil surface samples and measured peak resistance between 70 kPa to 195 kPa. The residual resistance found for those samples varied between 38 kPa to 170 kPa, with a sensitivity ranging from 1 to 3. The void ratios of these soils vary from 1.0 to 8.0, while their coefficients of compressibility (C_C) vary from 0.1 to 0.8. Regarding the compaction process, Hernández, *et al.* [2] report values of 65% optimum moisture and a dry unit weight of 9.6 kN/m³ in the drying path and 38% optimum moisture and a dry unit weight of 11.4 kN/m³ in the wetting path.

(a) (b)

Figure 2. SEM images of volcanic ash soil: (a) undisturbed, (b) compacted.

3. Discussion

Based on the previous physical characteristics, a discussion of this type of soil in road construction is raised. This discussion was presented from the physical properties of these soils and their application to road engineering regarding soil classification and compaction properties and quality assurance and control.

3.1. Soil classification

Soils from weathered volcanic ash in a hot-humid tropical environment are susceptible to changes in their properties because of remolding or oven drying procedures in the laboratory [3-5,7,22]. An erroneous soil classification may impact the appropriate definition of pavement design units or even imply unreal liquefaction susceptibility when silts or clays are classified as sands.

Figure 3, obtained by the authors, shows the change in grain size distribution caused by oven drying on volcanic ash soil from Manizales, Colombia. Based on the oven-dried sample, the soil would be classified as sand with some material finer than the 0.075 mm sieve. The actual soil should be classified as silt or clay, based on the laboratory's grain size analysis on samples without oven drying. Hernández, *et al.* [2] stated that volcanic ash soils have the characteristic that when clayey soils are hygroscopically dried and re-saturated, they are transformed into sandy soils.

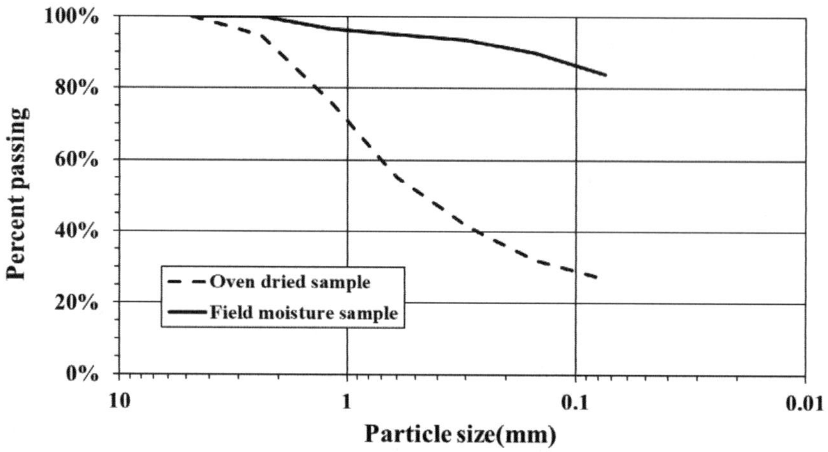

Figure 3. Comparison of grain-size analysis of a volcanic ash soil with and without oven drying from Manizales, Colombia.

Figure 4 shows the position in Casagrande's chart of several soils from the central Andean region of Colombia, including the city of Manizales, with data from [5] and [24]. All tests were performed on undried samples [25], and their approximate mineralogy corresponds to allophane or halloysite according to an approximate zonation of mineral composition presented in Wesley [17] and Fourie, *et al.* [26]. Díaz and Villada confirmed the presence of allophane, imogolite, and halloysite in the samples labeled as Manizales, Colombia, with X-Ray diffraction (XRD) and SEM [24].

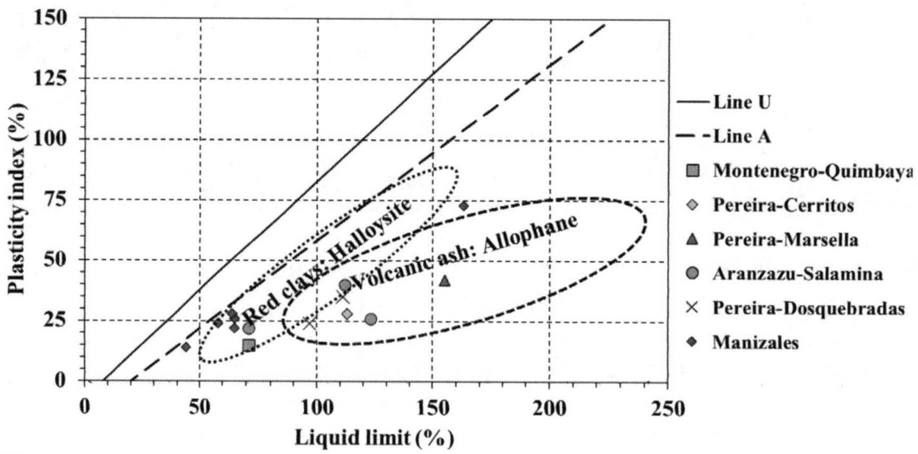

Figure 4. Position in Casagrande's chart of volcanic ash soils from the central Andean region in Colombia.

3.2. Compaction properties and quality assurance and control

Volcanic ash soils present two main challenges for compaction as a road construction material: first, as subgrade in cuts, the structure of the soil forbids the conventional compaction procedures with the sheep-roller compactor and other devices that cause large strains in the soil mass; second, in embankments, the compaction procedures require the use of low-pressure equipment (*i.e.*, wide-rail tractor selected by trial and error) and minimum handling from the borrow area to the extension and compaction at the construction site to avoid the disintegration of the soil.

The use of undisturbed samples is mandatory for assessing resistance in this type of soil. Thin-walled tubes provide appropriate samples for stability and deformation analysis; the in-situ sampling of California bearing ratio (CBR) mold [27] allows an approximate "undisturbed" technique for volcanic ash soils. The dynamic cone penetrometer (DCP) test [28] evaluates the undrained shear resistance of the subgrade with known correlations with the CBR and resilient modulus for subgrade evaluation [29-31]; for earthwork quality control construction, care must be taken in the laboratory estimation of compaction requirements based on the standard procedures to evaluate the unit weight-moisture-resistance relationships [32]. Figure 5 shows compaction curves for volcanic ash soils from two roads near Pereira, Colombia's central Andean region; both curves were obtained on samples with and without oven drying before the compaction in the laboratory.

In Figure 5(a), the in-situ condition (red diamond) corresponds to high moisture (95%) and a low dry unit-weight (6.835 kN/m³), which contrast with the characteristics of the oven-dried samples with a maximum dry unit weight of 10.2 kN/m³ and optimum moisture of 47%. Any attempt to construct an embankment based on the oven-dried compaction curve will require an impractical reduction of up to 50% of gravimetric moisture (several tons of water per truckload of borrowed soil) and an improbable increase of 146% in the dry unit weight. Figure 5(b) shows similar behavior; both cases show a clear difference between the optimum compaction moistures and the maximum dry unit weights of oven-dried and undried volcanic ash soils. Reference [4] reports the same behavior at Juan Santa Maria Airport construction in San José, Costa Rica, on residual tropical soils. Failure to identify the change in volcanic ash behavior due to laboratory practices as oven drying will cause a wrong assessment of the unit weight-moisture relationship. Such error will significantly impact construction activities as the quality control will be based on unrepresentative results. Any attempt to satisfy the low optimum moisture will require the setting-up of industrial drying installations with the associated costs, that is, to carry the alteration process from the small-scale samples in the lab to the whole earthwork construction at the field.

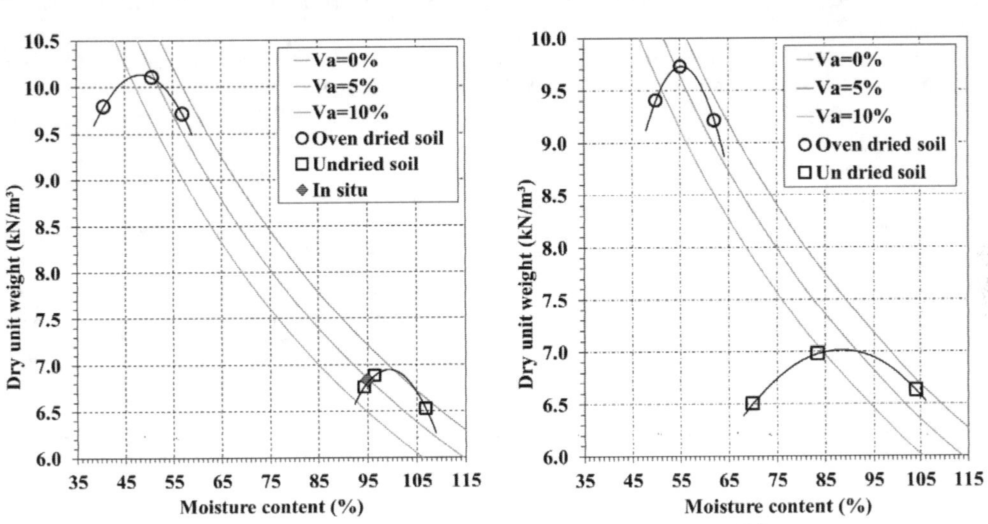

Figure 5. Compaction curves with oven-dried and undried samples obtained from (a) Pereira, Marsella, Colombia, K10+400 and (b) Pereira, Cerritos, Colombia, K1+700.

4. Conclusions

This study presents an understanding of the physical behavior of volcanic ash soils applied to engineering; volcanic ash soils formed in the hot-humid tropical environment have physical features that may be changed compared to the other soil in the world, these changes impact road construction activities. For example, the soil has a low density but stability in slopes with high angles. The compaction behavior changes with the wetting or drying path; these soils' unique behavior has been explained by their chemical composition, especially by the presence of minerals such as allophane, halloysite, and imogolite, which are susceptible to irreversible transformations in routine laboratory processes that were primarily conceived for sedimentary soils.

Two main aspects can impact the construction of roads in soils derived from volcanic ash in a tropical environment: the first is the error in classifying the soil as sand when it is silt or clay, which affects the definition of pavement design units and could even produce wrong conclusions about the susceptibility of the deposit to the liquefaction phenomena; the second impact compromises the validity of the reference tests for the quality control of the compaction of embankments, which represent unrealistic conditions in the field that will lead to the systematic rejection of batches, waste of resources, and delays because of the inadequate soil characterization. Since the design of embankments requires the definition of flow, deformation, and resistance parameters of the soil for stability and settlement analysis, the inadequate characterization of the compacted state also represents a risk for the users of such earth structures. These situations can be avoided with simple and inexpensive procedures for characterizing soil samples suspected of having these characteristics by comparing test results with drying and without prior drying of the samples before classification, resistance, and compaction testing in the laboratory.

Acknowledgments

The authors are grateful for the support for the elaboration of this work from the Universidad Nacional de Colombia.

References

[1] Lizcano A, Herrera M, Santamarina J 2018 Suelos derivados de cenizas volcánicas en Colombia *Revista Internacional de Desastres Naturales, Accidentes e Infraestructura Civil* **6(2)** 167

[2] Hernández O, Cordao-Neto M P, Caicedo B 2018 Structural features and hydro-mechanical behaviour of a compacted andesitic volcanic soil *Géotechnique Letters* **8(3)** 195

[3] Wesley L D 1973 Some basic engineering properties of halloysite and allophane clays in Java, Indonesia *Géotechnique* **23(4)** 471

[4] Morin W J, Todor P C 1970 *Laterite and Lateritic Soils and Other Problem Soils of the Tropics* (Rio de Janeiro: Road Research Institute, Brazilian National Highway Dept)

[5] Vásquez L C 1995 El Índice de Liquidez en la identificación de suelos sensibles a cambio de propiedades por el remoldeo y/o por el secado *Boletín de Vías* **22(82)** 33

[6] Betancur Y, Builes M, Millán A 2013 Variación de las propiedades mecánicas de arcillas alófanas en Colombia al variar el grado de saturación *Revista EIA* **10(20)** 173

[7] Herrera M C 2005 *Suelos Derivados de Cenizas Volcánicas en Colombia: Estudio Fundamental e Implicaciones en Ingeniería* (Bogotá: Universidad de Los Andes)

[8] Herrera M C, Lizcano A, Santamarina J C 2007 Colombian volcanic ash soils *Characterization and Engineering Properties of Natural Soils* ed Tan T, Phoon K K, Hight D W, Leroueil S (London: Taylor & Francis) p 2385

[9] Arango J D 1993 *Características Generales y Comportamiento Geotécnico de las Cenizas Volcánicas en la Zona del Antiguo Caldas* (Bogotá: Universidad de Los Andes)

[10] Bommer J, Rolo R, Mitroulia A, Berdousis P 2002 Geotechnical properties and seismic slope stability of volcanic soils *12th European Conference on Earthquake Engineering* (London: Elsevier)

[11] Forero-Dueñas C A, Gálvez-Peralta E, Fino W, Ulloa E F 1999 Estudios de la estructura de las cenizas volcánicas de Armenia y su relación con el comportamiento geotécnico *X Jornadas Geotécnicas de la Ingeniería Colombiana Boletín Colombiano de Geotecnia* (Bogotá: Sociedad Colombiana de Geotecnia)

[12] Yamin L E, Caicedo B, Solano E 1999 Evaluación de la respuesta dinámica del subsuelo en depósitos volcánicos y sus implicaciones para el diseño sísmico *Revista de ingeniería* **10** 79

[13] Mendoza C, Ramos R A, Calle O C 2021. Seismic evaluation of a profile of volcanic ash from the South American Andes *International Journal of Geotechnical Earthquake Engineering (IJGEE)* **12(1)** 40

[14] Shoji S, Nanzyo M, Dahlgren R 1993 *Volcanic Ash Soils: Genesis, Properties, and Utilization* (Amsterdam: Elsevier Science)

[15] Cerda A, Calvo A, Lavee H, Imeson A 1996 Erosionabilidad del suelo a lo largo del gradiente climático Coll de Rates - Benidorm Alacant *IV Reunión de Geomorfología* (A Coruña: Sociedad Española de Geomorfología)

[16] Terlien M 1997 Hydrological landslide triggering in ash-covered slopes of Manizales (Colombia) *Geomorphology* **20** 165

[17] Rendón M I, Viviescas J C, Osorio J P, Hernández M S 2020 Chemical, mineralogical and geotechnical index properties characterization of volcanic ash soils *Geotech. Geol. Eng.* **38** 3231

[18] García-Leal J C, Colmenares J E 2011 Predicción de la resistencia al corte en los suelos naturales derivados de ceniza volcánica *Pan-Am CGS Geotechnical Conference* (Toronto: ISSMGE)

[19] Garzón M, Colmenares Montañez J, Viveros Rosero L 2017 Influencia de la densidad seca y el contenido de agua de compactación en las curvas exploratorias de un suelo residual derivado de ceniza volcánica *INGE CUC* **13(1)** 19

[20] Malagón D, Pulido C, Llinas R, Chamorro C, Fernández J 1995 *Suelos de Colombia: Origen, Evolución, Clasificación, Distribución y Uso* (Bogotá: Instituto Geográfico Agustín Codazzi)

[21] Iyoda F, Hayashi S, Arakawa S, John B, Okamoto M, Hayashi H, Yuan G 2012 Synthesis and adsorption characteristics of hollow spherical allophane nano-particles *Applied Clay Science* **56** 77

[22] Wesley L D 2011 Geotechnical properties of residual soils *Boletín Colombiano de Geotecnia - XIII Congreso Colombiano de Geotecnia* (Bogotá: Sociedad Colombiana de Geotecnia)

[23] Moroto N 1991 Mechanical behaviour of two typical compacted volcanic soils in Hachinohe, Japan under different sample preparation methods *Soils and Foundations* **31(2)** 108

[24] Díaz E G, Villada C A 2002 Aplicación d*e la Difracción de Rayos X y Microscopía Electrónica para la Determinación de Minerales Detectables por estos Métodos en Suelos Derivados de Cenizas Volcánicas* (Manizales: Universidad Nacional de Colombia)

[25] American Society for Testing and Materials (ASTM) 2017 *Standard Test Methods for Liquid Limit, Plastic Limit, and Plasticity Index of Soils, ASTM D4318-17e1* (United States of America: American Society for Testing and Materials)

[26] Fourie A B, Irfan T Y, Queiroz de Carvalho J B, Simmons J V, Wesley L D 2012 Microstructure, mineralogy, and classification of residual soils *Mechanics of Residual Soils* ed Blight G E, Leong E C (Boca Raton: CRC Press) p 41

[27] American Society for Testing and Materials (ASTM) 2016 *Standard Test Method for California Bearing Ratio (CBR) of Laboratory-Compacted Soil, ASTM D1833-16* (United States of America: American Society for Testing and Materials)

[28] American Society for Testing and Materials (ASTM) 2018 *Standard Test Method for the use of the Dynamic Cone Penetrometer in Shallow Pavement Applications, ASTM D6951/D6951M-1* (United States of America: American Society for Testing and Materials)

[29] Zohrabi M, Scott P L 2003 *TRL Report TRL587 The Correlation Between the CBR value and the Penetrability of Pavement Construction Materials* (London: TRL Limited)

[30] Burnharm T, Johnson D 1993 *In Situ Foundation Characterization Using the Dynamic Cone Penetrometer"* (Maplewood: Minnesota Department of Transportation)

[31] Salgado R, Sungmin Y 2003 *FHWA/IN/JTRP-2002/30 - Dynamic Cone Penetration Test (DCPT) for Subgrade Assessment* (Indianapolis: Indiana Department of Transportation)

[32] American Society for Testing and Materials (ASTM) 2016 *Standard Test Methods for Laboratory Compaction Characteristics of Soil Using Standard Effort (12 400 ft-lbf/ft3 (600 kN-m/m3)), ASTM D698-12e2* (United States of America: American Society for Testing and Materials)

Journal of Physics: Conference Series **2118** (2021) 012013 doi:10.1088/1742-6596/2118/1/012013

Evaluation of mineral oils as matrices for AISI/SAE–1020 steel naphthenic corrosion study

G R Conde-Rodríguez[1], J A Sanabria-Cala[1,2,3], R A Mancilla Estupiñán[2], D A Laverde Cataño[1], and M C Núñez Castañeda[1]

[1] Grupo de Investigaciones en Corrosión, Universidad Industrial de Santander, Bucaramanga, Colombia

[2] Grupo Ambiental de Investigación Aplicada, Universidad de Santander, Bucaramanga, Colombia

[3] Laboratorio de Química, Colegio Integrado San Antonio, California, Colombia

E–mail: conderafaelg@gmail.com

Abstract. Petrochemical industry has suffered great economic impact due to light crude oil reserves reduction, so refineries have been processing high acidity heavy crude oils. Studies of corrosion caused by naphthenic acids are interfered by presence of other corrosive agents contained in real crude oils, so naphthenic phenomenon must be isolated using synthetic crude oils. For this reason, in present work two high purity mineral oils were used to evaluate their efficiency as synthetic crude oil matrices in AISI/SAE–1020 steel naphthenic corrosion study. Temperature levels evaluated were 200 °C, 250 °C, and 300 °C, while exposure times evaluated were 5, 10 and 15 hours. Surface morphological characterization of AISI/SAE–1020 steel was carried out using scanning electron microscopy and X–ray diffraction. Gravimetric tests showed that AISI/SAE–1020 steel naphthenic corrosion rate increases with temperature and exposure time for one of the synthetic crude oils. However, results obtained for the other synthetic crude oil did not show increasing behaviour due to presence of sulfur traces in the oil, which caused an interference with AISI/SAE–1020 steel naphthenic corrosion study, reducing the reliability of gravimetric results so they cannot be extrapolated to operating conditions in distillation units.

1. Introduction

In petrochemical industry, corrosion phenomenon is strongly associated with processing of heavy crude oil due to its high content of corrosive agents [1]. Naphthenic acids are among the main causes of this phenomenon, which are potentially corrosive and affect preheating furnaces, transfer lines and atmospheric and vacuum distillation units, causing fouling and decreasing the quality of middle distillates obtained [2].

Naphthenic corrosion is influenced by multiple variables such as temperature (generally occurs above 220 °C), exposure time, concentration and type of naphthenic acids and the presence of organic sulfur compounds, among others. This phenomenon is caused by reaction of naphthenic acids (RCOOH) with iron (Fe) from metallic surfaces, generating iron naphthenates ($Fe(RCOO)_2$) that are soluble in crude oil and cause a constant material loss, giving place to appearance of pitting corrosion in the exposed materials [3].

Currently, there are several alternatives for naphthenic corrosion prevention and mitigation such as: blending, which consists of mixing crude oil with others with lower naphthenic acids content to reduce the acidity; chemical neutralization of naphthenic acids; and metallurgy upgrade through selection of materials with greater corrosion resistance [3]. Evaluation of these methods is carried out in its initial stage through bench scale tests, generally by gravimetry, where fractions of real crude oil or synthetic crude oil can be used.

One advantage of using synthetic crude oils is that it allows to isolate naphthenic acids effect from other corrosive agents present in real crude oil fractions such as organic sulfur compounds, hydrochloric acid, amines and heavy metals, among others [4]. In addition to this, selection of mineral oil matrix is of great relevance during the development of bench scale tests, since its composition must be such that it does not generate any type of interference on the gravimetric results.

For this reason, in present work two mineral oils were used as synthetic crude oil matrices to evaluate the effect of temperature and exposure time on naphthenic corrosion of AISI/SAE–1020 steel in a system that simulates operating conditions of a transfer line in Colombian heavy crude processing.

2. Experimental

Tests were carried out using gravimetric coupons of AISI/SAE–1020 steel with dimensions: 76 mm length, 12.6 mm width, 1.5 mm thickness and 7.9 mm internal diameter for fastening hole. Surface metallographic preparation was carried out according to ASTM G1–03 standard guidelines [5]. Material surface was polished with silicon carbide abrasive paper No. 150, 220, 400, 600, 1500 and 2000; subsequently, gravimetric coupons were cleaned with acetone in ultrasonic bath and weighed on a Mettler AE 240 analytical balance with ± 0.0001 g precision. Chemical composition of AISI/SAE–1020 steel was determined by atomic emission spectrometry according to ASTM E415–15 standard [6] as shown in Table 1.

Table 1. Chemical composition (%wt) of AISI/SAE–1020 steel.

Element	C	Fe	Mn	S	Si	P	Ni	Mo
Composition (%wt)	0.17	98.61	0.82	<0.15	0.04	0.01	0.01	<0.01

Synthetic crude oils used were called SCO–1 and SCO–2, and they were prepared by adding a mixture of Aldrich analytical grade naphthenic acids on mineral oils MO–1 and MO–2, respectively. The amount of mineral oil and naphthenic acids were calculated to obtain synthetic crude oils with total acid number (TAN) equivalent to 1, which represents the amount of mg of potassium hydroxide needed to neutralize one gram of crude (mgKOH/g) [7].

Gravimetric tests were carried out in a Parr 4848 dynamic autoclave–type batch reactor with pressure and temperature control, where gravimetric coupons were placed immersed in 500 mL of the corresponding synthetic crude oil. Prior to start of each test, a purge with analytical grade nitrogen was carried out to remove oxygen dissolved in the synthetic crude oil and avoid interference from material oxidation reaction. Once the gravimetric test was completed, the coupons were removed from the reactor and weighed on an analytical balance to determine the mass loss, which was used to calculate the naphthenic corrosion rate of AISI/SAE–1020 steel according to Equation (1) [5], where: Δw represents mass loss in (g); A represents the exposed area in (cm^2); ρ represents the density of AISI/SAE–1020 steel in ($g\ cm^{-3}$); and t represents exposure time in (hours).

$$\text{Corrosion rate} = \frac{3.45 * 10^6 * \Delta w}{\rho * A * t}. \quad (1)$$

For each synthetic crude oil (SCO–1 and SCO–2) a 3^2 design of experiments was carried out, where levels established for temperature were 200 °C, 250 °C and 300 °C while levels for exposure time were 5, 10 and 15 hours. Each set of synthetic crude oil, temperature and exposure time conditions was evaluated in triplicate in order to guarantee the results repeatability. Surface morphological characterization of the exposed AISI/SAE–1020 steel was performed using scanning electron microscopy (SEM) combined with energy dispersive X–Ray spectroscopy (EDS) and X – ray diffraction (XRD).

3. Results and discussion

Figure 1 shows AISI/SAE–1020 steel corrosion rate exposed to SCO–1 and SCO–2 crude oils as a function of system temperature and exposure time, where is observed different behaviours for each synthetic crude oil.

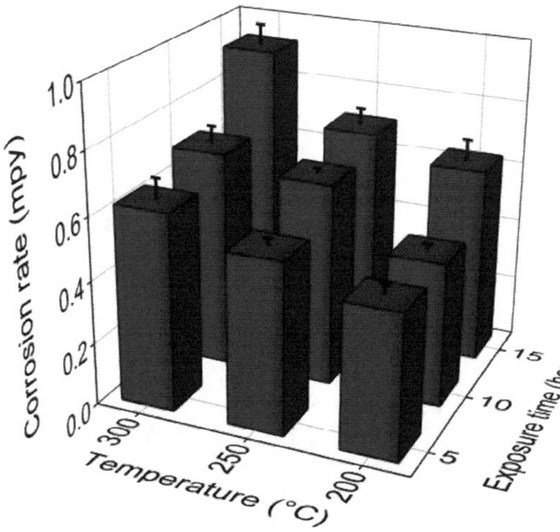

(a)

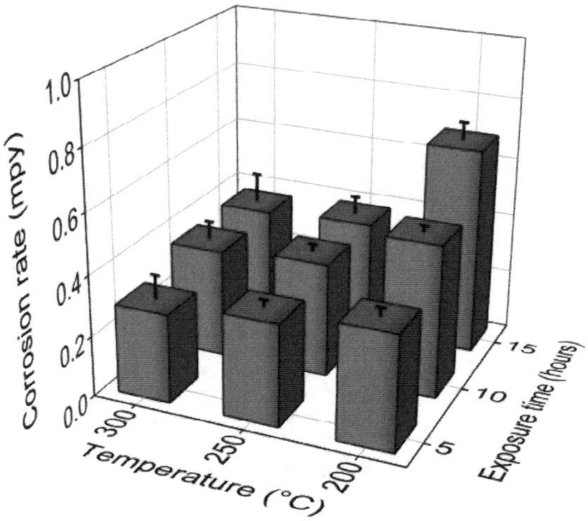

(b)

Figure 1. AISI/SAE–1020 steel corrosion rate for crude oil: (a) SCO–1 and (b) SCO–2.

For crude oil SCO–1, AISI/SAE–1020 steel corrosion rate increased with both temperature and exposure time. In the first case, phenomenon is associated with the greater reactivity of naphthenic acids when system temperature increases, since corrosion reaction occurs faster and allows forming a greater quantity of iron naphthenates, leading to an increase in AISI/SAE–1020 steel mass loss and corrosion rate [8]. In addition to this, iron naphthenates formed are soluble in crude oil, so they do not form a layer of corrosion products on the material surface that can isolate it from the corrosive environment, and for this reason the naphthenic corrosion reaction occurs continuously and progressively in the system, allowing the corrosion rate of the material to increase with exposure time [9]. In accordance with this trend, it is possible to infer that synthetic crude oil SCO–1 allowed to isolate the AISI/SAE–1020 steel corrosion rate values corresponding to naphthenic acids exposure, and therefore the mineral oil MO–1 can be used as a matrix in the study of naphthenic corrosion phenomenon.

On the other hand, behaviour of AISI/SAE–1020 steel corrosion rate for synthetic crude oil SCO–2 was different from that expected after exposure to a system where naphthenic phenomenon predominates. This was evidenced as a decreasing trend of corrosion rate with temperature for all exposure times, which is a behaviour commonly associated with sulfidic corrosion [10]. In this phenomenon, layers of corrosion products are formed on the material surface and act as a physical barrier that isolates the iron from the corrosive medium, thereby reducing the interaction between the metal surface and the naphthenic acids in the synthetic crude oil [11]. In this way, it is possible to infer that traces of sulfur were found in the composition of mineral oil MO–2 that reacted with AISI/SAE–1020 steel surface to form iron sulfides, and since these are insoluble in crude oil, they led to formation of adhered layers that prevented the advance of naphthenic corrosion reaction.

Figure 2 shows the surface morphological characterization of AISI/SAE–1020 steel exposed to synthetic crude oil SCO–1 at 300 °C for 15 hours, where the damage caused by naphthenic acids to the material is evidenced. As described in Equation (2), naphthenic corrosion reaction leads to formation of soluble iron naphthenates in crude oil, so the material presents an iron loss that is evidenced as an erosion corrosion attack on the metal surface [8].

$$Fe + 2\,RCOOH \rightarrow Fe(RCOO)_2 + H_2. \tag{2}$$

In addition to this, according to the hydrodynamic conditions of the system there is the possibility that areas where occurs a localized increase of naphthenic acids concentration are generated, as generally occurs in the plates of distillation units, which can lead to pitting corrosion that is difficult to identify on units in service and can cause serious material, economic, safety and environmental damage [12].

In the same way, lack of corrosion products on the material surface leads to the absence of major changes in elemental composition obtained through EDS analysis, where the iron from the material predominates and there are low concentrations of carbon and oxygen associated with both the original composition of the material and the thermal decomposition of naphthenic acids in the system. Finally, analysis highlights the absence of sulfur in the surface elemental composition, which confirms that mineral oil MO–1 oil was sulfur free and can be adequately used as a matrix for preparation of synthetic crude oils in naphthenic corrosion study.

Figure 3 shows the surface morphological characterization of AISI/SAE–1020 steel exposed to synthetic crude oil SCO–2 at 300 °C for 15 hours, where it is possible to observe the simultaneous presence of naphthenic and sulfidic corrosion, with areas with erosion corrosion attack caused by naphthenic acids and areas where iron sulfide–type corrosion products were formed.

As described for synthetic crude oil SCO–1, naphthenic corrosion is represented by Equation (2) where the iron in the material reacts with naphthenic acids to form iron naphthenates that dissolve in the crude and allow continuous attack to the metallic surface. However, for synthetic crude oil SCO–2, the presence of a high sulfur content was also evidenced in the elemental chemical composition of the exposed AISI/SAE–1020 steel. This was caused by the presence of sulfur traces in mineral oil MO–2, generally organic sulfur compounds, which thermally decompose to generate hydrogen sulfide that reacts with the metal surface and forms iron sulfides as shown in Equation (3) [13].

$$Fe + H_2S \rightarrow FeS + H_2. \tag{3}$$

Element	Carbon	Oxygen	Iron
%wt	1.65	2.14	96.21

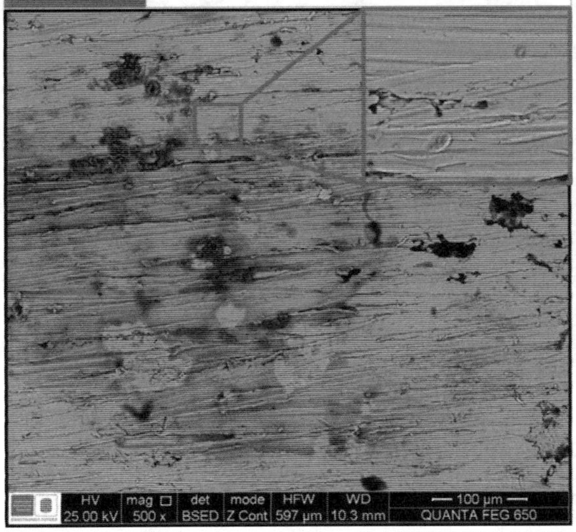

Figure 2. SEM micrographs of AISI/SAE–1020 steel exposed to crude oil SCO–1.

Element	Carbon	Oxygen	Sulfur	Iron
%wt	1.71	2.26	25.16	70.87

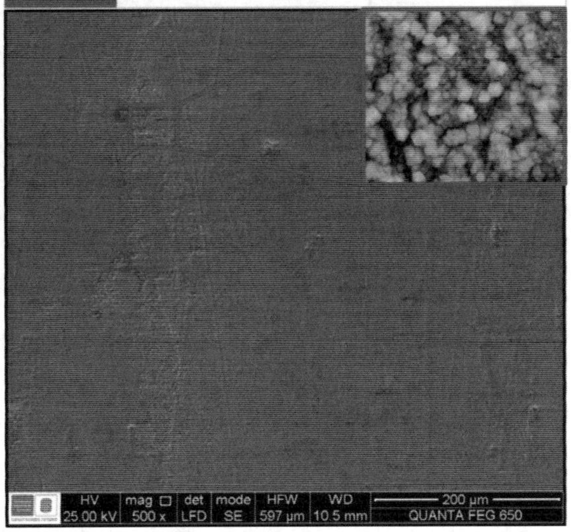

Figure 3. SEM micrographs of AISI/SAE–1020 steel exposed to crude oil SCO–2.

Iron sulphides formed as corrosion products are insoluble in crude oil, which is why they remain attached to the surface and generate a mass gain that distorts the gravimetric results. In this way, the mass loss used to determine AISI/SAE–1020 steel corrosion rate does not correspond solely to naphthenic phenomenon, but is interfered by mass gain caused by sulfidic phenomenon. Additionally, the increase in system temperature favours the thermal decomposition of sulfur compounds, increasing the iron sulfides formation and, therefore, reducing the corrosion rate of the AISI/SAE–1020 steel, as observed in gravimetric results [14].

Additionally, it is possible that a third chemical reaction occurs as presented in Equation (4), where interaction between naphthenic acids and the iron sulfides formed is observed, increasing the system complexity [15]. In this way, the presence of traces of sulfur in mineral oil MO–2 generates a great disturbance in the system, and for this reason it should not be used as a matrix for the preparation of synthetic crude oils in naphthenic corrosion study.

$$FeS + 2\,RCOOH \rightarrow Fe(RCOO)_2 + H_2S. \tag{4}$$

Finally, Figure 4 shows the X–Ray Diffraction patterns for AISI/SAE–1020 steel exposed to 300 °C for 15 hours. For synthetic crude oil SCO–1, the presence of only a high intensity peak close to 45 ° is evidenced, corresponding to the iron in the metallic matrix. However, for synthetic crude oil SCO–2 the additional presence of an intense peak at 43° and multiple peaks between 30° and 35° is evidenced, which correspond to iron sulfides formed on the steel surface due to sulfidic corrosion, confirming the results obtained in gravimetric and microscopy analysis [16].

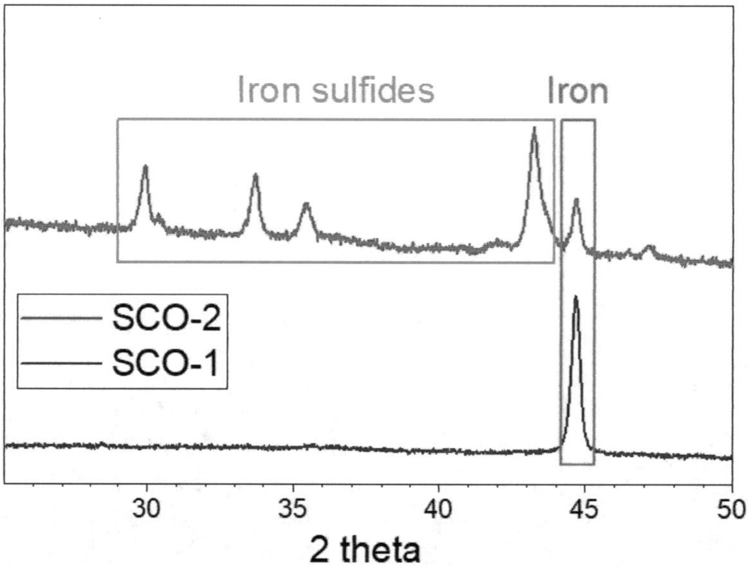

Figure 4. XRD patterns of AISI/SAE–1020 steel exposed at 300 °C for 15 hours.

4. Conclusions

Naphthenic corrosion rate of AISI/SAE–1020 steel exposed to synthetic crude oil SCO–1 increased with the two operating variables analysed, since the increase in temperature caused a greater reactivity of the naphthenic acids to form iron naphthenates, which did not form a physical barrier that isolated the material from the corrosive environment and therefore the reaction took place continuously over time.

On the contrary, for synthetic crude oil SCO–2 a great difference was evidenced for AISI/SAE–1020 steel corrosion rate trend with the system temperature, due to the presence in the mineral oil of sulfur traces that allowed the formation of iron sulfides on the material surface, increasing the mass of the system and interfering with gravimetric results.

For this reason, gravimetric tests oriented to evaluate naphthenic corrosion that are carried out using synthetic crude oils must start from sulfur–free oil matrices, in such a way that reliable studies of naphthenic phenomenon are obtained, which are very useful for advance and optimization of corrosion monitoring and control systems in the petrochemical industry.

Acknowledgments

The authors want to thank Universidad Industrial de Santander and Universidad de Santander, Colombia, for its technical and financial support during the development of this research project.

References

[1] Mejía C, Laverde D, Molina D 2015 Correlation for predicting corrosivity of crude oils using proton nuclear magnetic resonance and chemometric methods *Energy Fuels* **29** 7595

[2] Slavcheva E, Shone B, Tumbull A 1999 Review of naphthenic acid corrosion in oil refining *British Corrosion Journal* **34** 125

[3] Tumbull A, Slavcheva E, Shone B 1998 Factors controlling naphthenic acid corrosion *Corrosion* **54** 922

[4] Freitas S, Malacarne M, Romão W, Dalmaschio G, Castro E, Celante V, Freitas M 2013 Analysis of the heavy oil distillation cuts corrosion by electrospray ionization FT–ICR mass spectrometry, electrochemical impedance spectroscopy, and scanning electron microscopy *Applied Surface Science* **104** 656

[5] American Society for Testing and Materials (ASTM) 2003 *Standard Practice for Preparing, Cleaning, and Evaluating Corrosion Test Specimens, ASTM G1–03* (United State of America: American Society for Testing and Materials)

[6] American Society for Testing and Materials (ASTM) 2015 *Standard Test Method for Analysis of Carbon and Low-Alloy Steel by Spark Atomic Emission Spectrometry, ASTM E415–15* (United State of America: American Society for Testing and Materials)

[7] Flego C, Galasso L, Montanari L, Gennaro M 2014 Evolution of naphthenic acids during the corrosion process *Energy Fuels* **28** 1701

[8] Patrick B 2015 *Understanding Naphthenic Acid Corrosion in Refinery Settings* (Berkeley: University of California)

[9] Qu D, Zheng Y, Jing H, Yao Z, Ke W 2006 High temperature naphthenic acid corrosion of Q235 and 5Cr1/2Mo steels in synthetic refining media *Corrosion Science* **48** 1960

[10] Dettman D, Li N, Wickramasinghe D, Xu Z, Chen X, Elliott G, Luo J 2012 The influence of naphthenic acid and sulfur compound structure on global crude corrosivity under vacuum distillation conditions *Corrosion NACE* **1326** 120045

[11] Jin P, Nesic S, Wolf A 2015 Analysis of corrosion scales formed on steel at high temperatures in hydrocarbons containing model naphthenic acids and sulfur compounds *Surface and Interface Analysis* **47** 454

[12] Kaneko M, Isaacs H 2002 Effects of molybdenum on the pitting of ferritic –and austenitic– stainless steels in bromide and chloride solutions *Corrosion Science* **44** 1825

[13] Rebak R 2011 Sulfidic corrosion in refineries – a review *Corrosion Reviews* **29** 123

[14] Lepore J 2016 *The role of sulfur Species in Establishing the Corrosion Reactions in Refinery Metallurgies* (Edmonton: University of Alberta)

[15] Huang B, Yin W, Sang D, Jiang Z 2012 Synergy effect of naphthenic acid corrosion and sulfur corrosion in crude oil distillation unit *Applied Surface Science* **259** 664

[16] Francis H 2015 *Mechanisms Governing the Growth, Reactivity and Stability of Iron Sulfides* (Cambridge: Massachusetts Institute of Technology)

Durability assessment of concretes reinforced with steel and synthetic fibers in Colombia

J Carrillo[1], N Guerrero[2], O Arroyo[3], and A Rodríguez[1]
[1] Departamento de Ingeniería Civil, Universidad Militar Nueva Granada, Colombia
[2] Departamento de Ingeniería Civil, Universidad de Ibagué, Ibagué, Colombia
[3] Departamento de Ingeniería Civil, Universidad de La Sabana, Chía, Colombia

E-mail: julian.carrillo@unimilitar.edu.co

Abstract. In recent years, the use of steel fibers has increased in Colombia, optimizing structure designs, quality specifications and mechanical behavior. However, the concrete durability and resistance to aggressive environments such as sulfates and chlorides are not commonly evaluated. This paper includes results of durability analysis for 21 MPa compressive strength concretes reinforces with steel and synthetic fibers. Performance parameters measured during tests include sulfate resistance, chlorides penetration, and carbonation according to technical standards. The Variables of experimental program were fiber type and fiber dosage. Dosages for synthetic fibers were 1.5 kg/m^3, 3.0 kg/m^3 and 5.0 kg/m^3, and for steel fibers were 5.0 kg/m^3, 9.0 kg/m^3 and 18 kg/m^3. Results of the study were used to propose recommendations for reducing concrete deterioration when exposed to these aggressive environments and for extending the durability service life.

1. Introduction

Currently, the term durability includes fundamental aspects to obtain an ideal concrete, that is described in American Concrete Institute Standard (ACI), ACI 318R-19 [1], the most important are exhibition and resistance by the action of environmental aggressive agents (sulfates, chlorides (CO_2) etc.) [2,3].
It is well known that the fiber-reinforced concrete (FRC) improved plastic retraction characteristics; also increases strength tension and energy absorption, decrease thicknesses of fissures in specimen reducing permeability. These mechanical characteristics of the FRC can improve construction processes that cannot be obtained by reinforcement bars or welded wire mesh, but even like that it is not achieved to measure the impact that could have the structure exposed normally before the action of some aggressive agent [4-6].

Colombia is in a great process of development, and the civil engineering has turned into one of their engines to achieve it. The implementation of new technologies of construction like the use of fibers of steel or synthetic fibers in the fiber-reinforced concrete, which achieves resistance to aggressive environments and mechanical ideal properties, offers to the engineer a new perspective of new designs. In this paper the assessment of use of fibers in concrete specimen subjected to aggressive agents is accomplished.

2. Experimental program

The fiber-reinforced concrete durability evaluation with steel and synthetic fiber was realized by means of the test of 66 specimens by 3 dosages for fiber type and 11 specimens without fiber as comparison standard.

2.1. Characteristics of the specimens

In the experimental program there was included durability tests as ion chloride resistance, and ion sulfate resistance. Additional mechanical tests were done to evaluate the compressive strength. In different specimen there was had variation of steel and synthetic dosages fibers. The dosages assigned are according to the Colombian standard, "Normas sismo resistentes (NSR), NSR-10", [7] which allows the use of 60 kg/m³ fibers for girders as dosage, but not for other types of concrete elements. For this reason, this investigation chose dosage to 18 kg/m³ for industrial floor slabs, and it was reduced proportionally to the half for both types of fibers as is specified in the Table 1. Each specimen was named in agreement to its content of fibers (type and dosage) as 1-2, where 1 indicates the type of fiber (S = synthetic, A = steel) and 2 indicates the dosage used in kg/m³. In specimen without fiber, the classification assigned were the letter T. The different cylinders were tested to evaluate durability to an age of 28 days; to evaluate compressive strength, specimens were tested to 3, 14 and 28 days. The quantity of tested specimens by ages is shown in Table 2.

Table 1. Study variables.

Variable		Description			
Type of fiber	Steel	Length (lf) mm			50.0
		Diameter (df) mm			1.1
		Aspect relation			45
	Synthetic	Length (lf) mm			5.5
		Thickness (ef) mm			0.3
		Aspect relation			18.3
Dosages for type of fiber	Steel	Df, Kg/m³	1.50	3.00	5.00
		Volume fraction (Vf) %	0.15	0.30	0.51
	Synthetic	Df, Kg/m³	5.00	9.00	18.00
		Volume fraction (Vf) %	0.51	0.91	1.83

2.2. Concrete properties

The nominal concrete compressive strength (f'c) was 21 MPa, the maximum nominal aggregate size was 1 inch and minimal slump was 7 inches. In addition, the water/cement relation (W/C) was 0.57 and the relation of sand and gravel was 80% and 20% respectively. The specimens were processed in a special room with temperature not bigger than 25 °C and continuous humidity.

2.3. Test setup

All mechanical and durability tests were realized in the Argos Colombia Laboratory's, Puente Aranda Plant, located in Bogota, Colombia.

2.3.1. Tests in fresh condition. In every dosage, the slump concrete test was done according to "Norma Técnica Colombiana (NTC), NTC-393" [8]. In agreement to European standards in Spanish, UNE-EN 14488-7 [9], which determines the real dosing of fibers by extraction in 3 cylinders per dosage, the procedure was realized after spilling the concrete in the mold to prevent which will act the cementitious paste. The real dosage of fibers was calculated from the volume of the cylinder and the weight of the fibers extracted. Additional tests were done to measure the unit weight in fresh condition (M), and the air content (A), of fiber-reinforced concrete, according to NTC-1926 [10].

Table 2. Test summary FRC (steel and synthetic).

Test type	Age, days																				
	3							7							28						
	T	A-5	A-9	A-18	S-1.5	S-3	S-5	T	A-5	A-9	A-18	S-1.5	S-3	S-5	T	A-5	A-9	A-18	S-1.5	S-3	S-5
Resistance to Strength compressive	2	2	2	2	2	2	2	2	2	2	2	2	2	2	2	2	2	2	2	2	2
Resistance ion chloride															1	1	1	1	1	1	1
Nord test															1	1	1	1	1	1	1

2.3.2. Compressive strength test. Compressive strength tests in FRC (steel and synthetic), were done according to guidelines of NTC-673 [11]. The tests were carried out in a universal machine MATEST with a maximum capacity of 2000 kN; to ensure uniform load distribution to specimen, sheets of neoprene were used according to NTC-3708 [12]. All mechanical and durability tests were carried out in the Argos Colombia Laboratory's- Puente Aranda Plant, located in Bogota, Colombia.

2.3.3. Chloride ion penetration resistance test. The chloride ion penetration resistance test in fiber-reinforced concrete were done according to American Association of State Highway and Transportation Officials (AASHTO), AASHTO T277, or American Society for Testing and Materials (ASTM), ASTM C1202 [13]. The chemist compounds were NaOH to 0.3 N and NaCl to 3%. The obtained result is the quantity of energy used for the transfer of substances, in Coulombs.

2.3.4. Sulfate ion penetration resistance test; Nord test. The sulfate ion penetration resistance tests were done according to Nordtest, NT BUILD 492 [14]. The setup is similar to chloride ion penetration resistance test, the equipment, cells, specimens sizes used is the same but chemical compounds used in the ionic transfer are different since there is used $Ca(OH)_2$ replacing the NaOH. In addition, the voltage was constant of 30 V and the measured result is the penetration in millimeters. Details of the experimental program are reported by Garzón [15].

3. Results and discussion

In order to evaluate appropriately the experimental obtained results, this study apply the parameters of classification stipulated in each procedure mentioned previously.

3.1. Fresh concrete properties

The actual specimen dosage obtained for two types of fibers is shown in Table 3. In addition, the measured values for steel fiber of slump, air content, unit weight, are shown in the Figure 1 and Figure 2 for synthetic fiber. In the Figure 1(a) can be observed a softly decrease of slump with addition of steel fiber. An increase of approximately 20 % of air content (A) is observed with addition of steel fibers (Figure 1(b)). The use of steel fibers does not affect specimens' unit weight as shown in Figure 1(c). Also, it can be observed in Figure 1(d) that actual dosage values are over the design content of fiber.

Table 3. Actual specimen dosages of steel and synthetic fiber.

Specimen	Design dosage (kg/m^3)	Actual dosage (kg/m^3)
A-5	5.0	5.5
A-9	9.0	9.7
A-18	18.0	16.4
S-1.5	1.5	1.7
S-3	3.0	3.7
S-5	5.0	6.4

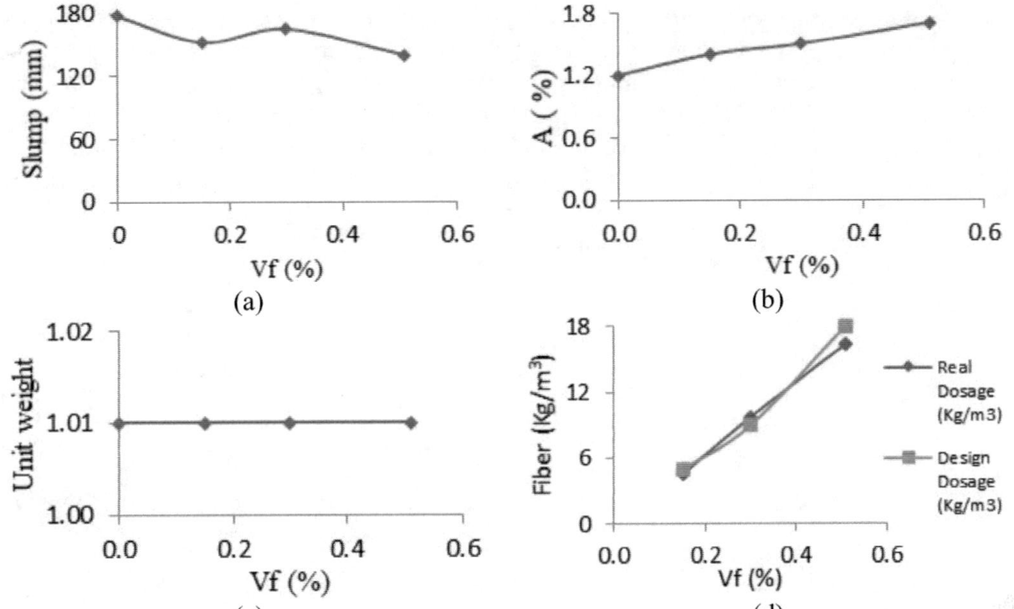

Figure 1. Fresh concrete properties for steel fiber; (a) slump, (b) air content, (c) unit weight, and (d) fiber content.

As can be observed in Figure 2(a), slump curve presents a significant decrease with increase of fiber because the fibers try to keep all the components of the concrete one close. An increase of air content (A) of approximately 30% is observed in Figure 2(b). A slightly change of unit weight it can be observed in Figure 2(c). For the case of the content of fibers, all actual values of dosage were over the values of design dosage, up to 28% for S-5 Figure 2(d).

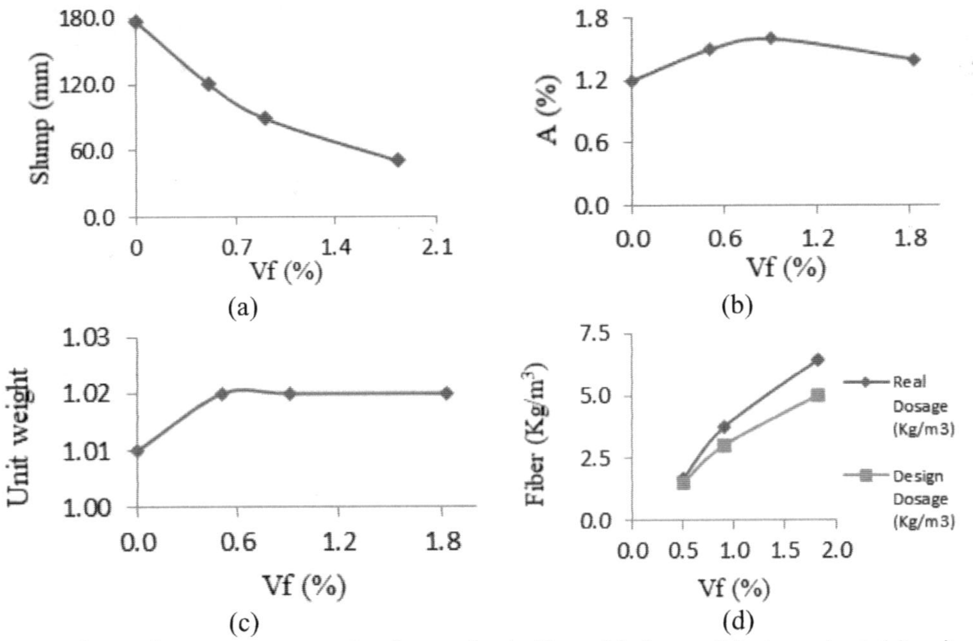

Figure2. Fresh concrete properties for synthetic fiber; (a) slump, (b) air content, (c) unit weight, and (d) fiber content.

3.2. Compressive strength properties

Figure 3 presents compressive strength curves of the fiber-reinforced concrete with steel fiber and synthetic fiber. It can be observed that all compressive strength results at 28 days of fiber-reinforced concrete are below the concrete simple. This result can be due to the distribution of the fiber in the specimen, which believes emptiness's between the fiber and the components of the concrete.

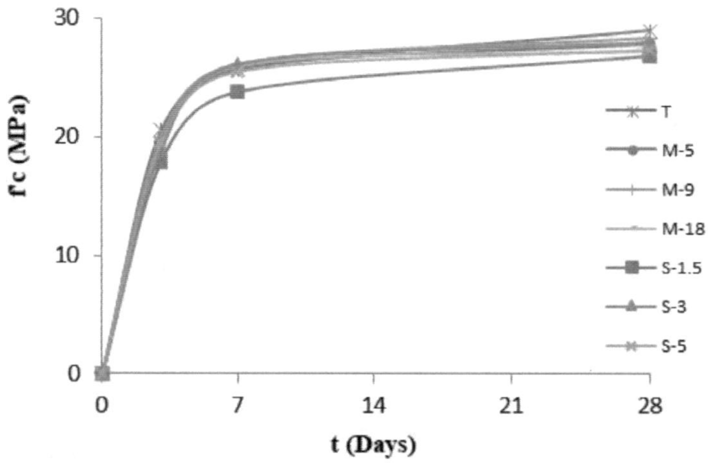

Figure 3. Evolution of strength compressive.

3.3. Chloride ion penetration properties

According to Figure 4 and based on the classification of Table 4, it shows that smaller steel fiber A-5 dosage and dosage of synthetic fiber S-5 exhibit lower loads chloride ion penetration Coulombs. The highest score has the specimen A-18, which is approximately 55% to concrete without fiber and has a high permeability, also the specimen S-3, as classified by ASTM C1202 [13].

Table 4. Classification results and chloride ion penetration resistance.

Specimen	Load Coulombs (C)	Classification according to ASTM-C1202 [13]
T	4001.5	High
A-5	1554.0	Low
A-9	3390.1	Moderate
A-18	6189.3	High
S-1.5	3788.1	Moderate
S-3	4004.5	High
S-5	2183.0	Low

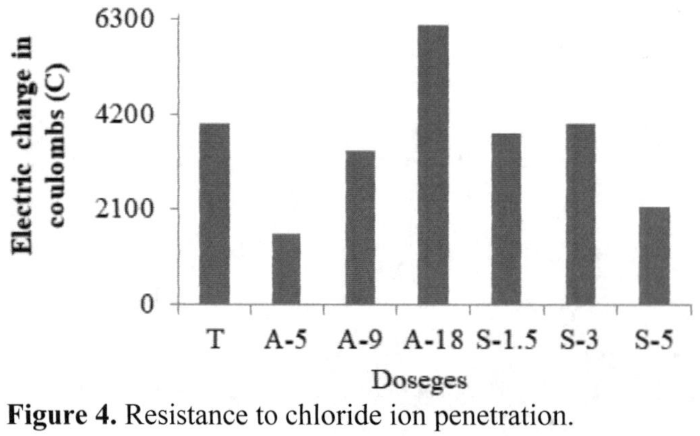

Figure 4. Resistance to chloride ion penetration.

3.4. Nord test properties

As shown in Table 5 the specimen with the highest average penetration was that one without dosing fiber (T), which is important because the change with the addition of fiber, evidenced by optimizing the conditions of durability and reducing the average penetration. It is clear that the specimen S-1.5 is the minimum value of sulfate ion penetration; it was about 93% lower than the highest score recorded. The best performance for the type of fiber, synthetic fiber has, as it has smaller penetration values regarding the results of the specimens of steel fiber.

Table 5. Nord test results test.

Specimen	Penetration (mm)
T	53
A-5	52
A-9	50
A-18	40
S-1.5	4
S-3	46
S-5	46

4. Conclusions

In this paper are presented the results of an experimental program to evaluate the durability of fiber-reinforced concrete with steel fiber and synthetic fiber. Addition of fiber in concrete generates a confinement effect by increasing the fiber content, the latter results in a decrease of penetration of water; this deduction is made by comparing the results of the specimens of each type of fiber with the specimen without fiber. Values of air content are increased with the inclusion of fiber in the concrete, reducing the gaps that were presented and optimizing the mix; all compressive strength test results of steel fiber reinforced, and synthetic fiber reinforced specimens were lower compressive strength without fiber specimen. The results are presented in terms of volume fraction in order to provide professional analysis and compare with other studies.

One of the main limitations of the study were data at different ages in all trials, so it could not perform numerical correlations or propose models to predict the impact of any aggressive agents that exposed concrete. Although the results obtained experimentally for the compressive strength of fiber-reinforced specimens are lower than those of unreinforced specimens, further research on concrete durability should be carried out, as it will allow the construction of structures that require less spending on maintenance and repair. In Colombia, conducting this type of research is important to encourage the use of fibers in concrete structures and optimize them to avoid having only a reactive behavior to pathologies in concrete structures. This is especially important for large infrastructure or housing developments that impact the quality of life of entire communities.

Acknowledgments

The authors gratefully acknowledge “Vicerrectoría de Investigaciones” at the Universidad Militar Nueva Granada (UMNG), Colombia, for financing the research project EXT-INV-3274. The donation of concrete and steel fibers from Argos, Colombia (Diego Velandia) and Proalco, Dramix (Liliana Cardona), respectively, is greatly appreciated. The authors also thank engineer Diego Garzón for his support during data and figures processing.

References

[1] American Concrete Institute (ACI) 2019 *Code Requirements for Structural Concrete and Commentary on Building Code Requirements for Structural Concrete, ACI 318R-19* (United States of America: American Concrete Institute)

[2] Aguirre A M, Mejía de Gutiérrez R 2013 Durability of reinforced concrete exposed to aggressive conditions *Materiales de Construcción* **63(309)** 7

[3] Hezhi Liu, Qian Zhang, Victor Li, Huaizhi Su, Chongshi Gu 2017 Durability study on engineered cementitious composites (ECC) under sulfate and chloride environment *Construction and Building Materials* **133** 171

[4] Singh H 2017 *Steel Fiber Reinforced Concrete. Behavior, Modelling and Design* (Singapore: Springer)

[5] Yuanxun Z, Xiaolong W, Guangxian H, Qingfang S, Jianguo X 2018 Mechanical properties of steel fiber-reinforced concrete by vibratory mixing technology *Advances in Civil Engineering* **2018** 1

[6] Chen Y, Cen G, Cui Y 2018 Comparative study on the effect of synthetic fiber on the preparation and durability of airport pavement concrete *Construction and Building Materials* **184** 34

[7] Asociación Colombiana de Ingeniería Sísmica (AIS) 2010 *Reglamento Colombiano de Construcción Sismo Resistente, NSR-10* (Colombia: Asociación Colombiana de Ingeniería Sísmica)

[8] Instituto Colombiano de Normas Técnicas y Certificación (ICONTEC) 1992 *Método de Ensayo para Determinar el Asentamiento del Concreto, Norma Técnica Colombiana, NTC-396* (Colombia: Instituto Colombiano de Normas Técnicas y Certificación)

[9] Asociación Española de Normalización y Certificación 2007 *Contenido de Fibras del Hormigón Reforzado con Fibras, UNE-EN 14488-7* (España: Asociación Española de Normalización y Certificación)

[10] Instituto Colombiano de Normas Técnicas y Certificación (ICONTEC) 2013 *Concretos. Método de Ensayo para Determinar la Densidad (Masa Unitaria), el Rendimiento y el Contenido de Aire por Gravimetría del Concreto, Norma Técnica Colombiana, NTC-1926* (Colombia: Instituto Colombiano de Normas Técnicas y Certificación)

[11] Instituto Colombiano de Normas Técnicas y Certificación (ICONTEC) 2010 *Concretos. Ensayo Resistencia a la Compresión de Especímenes Cilíndricos de Concreto, Norma Técnica Colombiana, NTC-673* (Colombia: Instituto Colombiano de Normas Técnicas y Certificación)

[12] Instituto Colombiano de Normas Técnicas y Certificación (ICONTEC) 2002 *Uso de Refrentado no Adherido para la Determinación de la Resistencia a la Compresión de Cilindros de Concreto Endurecido, Norma Técnica Colombiana NTC-3708* (Colombia: Instituto Colombiano de Normas Técnicas y Certificación)

[13] American Association of State Highway and Transportation Officials (AASHTO) 2012 *Standard Test Method for Electrical Indication of Concrete's Ability to Resist Chloride Ion Penetration, - USE ASTM C1202, AASHTO T277* (Washington, DC: American Association of State and Highway Transportation Officials)

[14] Nordtest 1999 *Concrete, Mortar and Cement-Based Repair Materials. Chloride Migration Coefficient from Non-Steady-State Migration Experiments, NT BUILD 492* (Finland: Nordtest)

[15] Garzón D 2016 Evaluation of durability properties of concrete reinforced with steel and synthetic fibers (Bogotá: Universidad Militar Nueva Granada)

Journal of Physics: Conference Series **2118** (2021) 012015 doi:10.1088/1742-6596/2118/1/012015

Scientific insights of electrochemical replacement approach to describe the origin of Pre-Columbian Peruvian gilded copper-based objects

A Alviz-Meza[1], D A Chipoco Haro[1,2], and M I Velarde[1]
[1] Centro de Investigación para la Conservación del Patrimonio, Universidad de Ingeniería y Tecnología, Lima, Perú
[2] Departamento de Ingeniería Química, Universidad de Ingeniería y Tecnología, Lima, Perú

E-mail: centropatrimonio1@utec.edu.pe

Abstract. Pre-Columbian Peruvian goldsmiths developed gilded copper-based objects by ancient techniques that require identification to propose conservation strategies. Lechtman H, conducted experiments to suggest that the electrochemical replacement was the gilding technique used by the Moche and Vicus cultural groups. Despite her remarkable achievement, the quantitative data provided by her is still open to discussion. This work focused on obtaining experimental data to recreate her protocol by introducing less gold precursor. Polished copper pieces were plated with an adherent gold film of up to 7.5 μm after immersing them into an electrolytic solution for 3 min and 6 min at 80 °C. Our results demonstrated that the electrochemical replacement technique gives rise to anodic regions in the plated objects. Further studies around the corrosion process that undergoes these heritage objects in burial and environmental conditions are suggested to determine their deterioration rate. Moreover, electroless and galvanic techniques should be explored in order to improve current approaches.

1. Introduction

Through the centuries that preceded the Spanish conquest, ancient Peruvian metalworkers developed various gilding techniques to change the metal surfaces of an object into different hues of gold and silver [1]. Those gilded objects were part of elite costumes and took part in the ritual paraphernalia that expressed and consolidated the power of rulers and priests. Most of the gilded copper objects were recovered from burial sites and are now part of museum collections. Despite their care, they are still undergoing severe corrosion processes that put their integrity at risk. Therefore, understanding the gilding techniques is not only essential to enrich our knowledge of the Andean metallurgy development, but more importantly, it provides scientific grounds to the practice of adequate conservation strategies that ensure the preservation of our metal cultural heritage [2].

Among the limited research done in this field, Lechtman H [3-5] was the first to identify and recreate the use of electrochemical replacement as a gilding technique, shared by the Moche and Vicus cultural groups from the north coast of Peru during the early intermediate period (100-500 AD). In her experimental work, she proposed using a mixture of KNO_3, $KAl(SO_4)_2 \cdot 12H_2O$, and NaCl to dissolve the gold. Then, the solution was neutralized with $NaHCO_3$ until a pH of 9 to gently deposit gold on copper objects. However, her published experimental results lacked the quantitative data related to the exact temperature and concentration of $NaHCO_3$. She reported a thin and brittle layer of gold formed on a cleaned copper substrate after dipping it into the bath for a few minutes. Thus, further annealing was required to create a strong bond between the layer and substrate.

Therefore, this research work focused on obtaining the missing experimental data to recreate Lechtman's protocol; for this, we studied the deposited gold on the copper substrate by X-ray fluorescence (XRF) while the uniformity of the layer was evaluated by optical microscopy. In addition, scanning electron microscopy (SEM) with energy-dispersive x-ray spectroscopy (EDS) was included to argue about the plating technique used by the Pre-Columbian goldsmiths from Dos Cabezas, North coast of Perú.

2. Methodology

An electrolytic replacement technique was developed based on Lechtman's work [3]. The methodological steps introduced are described below.

2.1. Copper coupons preparation

Copper coupons were cut in rectangles of 5.0 cm length x 2.5 cm height x 0.1 cm width, sanded sequentially with the silicon carbide (SiC) grit sizes of 500 μm, 800 μm, 1200 μm, and 2000 μm, and polished to a mirror-like finish through a 1.0 μm alumina powder slurry. Before deposition, each copper coupon was cleaned using ethanol (99.9%, Merck) and acetone (99.8%, Merck) in an ultrasonic bath (WUC-A01H, Witeg) for 5 min, respectively.

2.2. Electrolytic bath and gold deposition

The electrolytic bath was prepared in a 150 mL beaker containing 0.01 M of $HAuCl_4$ (99.995%, Sigma-Aldrich) as the gold precursor, 0.046 M of KNO_3 (99%. EMSURE), 0.01 M of $KAl(SO_4)_2 \cdot 12H_2O$ (A.C.S. Reagent, J.T. Baker), and 0.08 M of NaCl (100%, J.T. Baker) in 60 mL of distilled water. Then, 0.13 M of $NaHCO_3$ (99%, Merck) was added to reach a pH level above 9 (between 9.1 and 9.3). The prepared bath was stirred with a magnetic pill for 1.5 h at 80 °C. Finally, half of the copper specimen was dipped in the electrolytic bath. The copper coupon was lifted 1.5 cm after 3 min of reaction and removed from the solution after 6 min of testing.

2.3. Characterization of samples

The characterization of gilded copper coupons was performed through XRF and optical microscopy; besides, SEM-EDS analysis aimed to complement the results obtained in the lab by comparing them with the data provided by an archaeological Moche sample from Dos Cabezas, North of Perú.

2.3.1. X-ray fluorescence spectroscopy. These measurements were made through the Hitachi X-MET 8000 Expert equipment. The gilded copper was placed in the spot region of the equipment -avoiding anodic zones- for 20 s of measuring time in the alloy mode. The resulting spectra were reviewed to confirm the presence of gold in the sample. The obtained results correspond to the average of the two measurements made in both frontal faces of the coupons.

2.3.2. Scanning electron microscopy with energy-dispersive x-ray spectroscopy. This characterization was performed using a TESCAN second-generation SEM, VEGA/XMU model. The gilded copper samples were mounted in resin (TramsOptic) for metallographic samples (ZXQ-5A, LYRICTEST) and polished until mirror finishing before characterization.

2.3.3. Optical microscopy. The microscopic features of the sample were characterized after the gilding process through a continuous zoom trinocular stereo microscope (Alpha Optics, GL6545TI). The images were taken by a trinocular camera of 18 megapixels (Amscope software).

3. Results and discussion

Heather Lechtman proposed an effective method for gilding copper with a corrosive bath rich in reagents possible known and available for ancient Peruvian Piura Valley citizens [3-5]. We took the same 1:1:1 mass ratio of potassium aluminum sulfate, potassium nitrate, and sodium chloride that she reported. The mass required per reagent in the solution was set based on the sodium chloride molecular weight -the lowest-. Then, according to a similar study [6], we selected the specific concentration of 0.08 M for the sodium chloride. Meanwhile, for the gold precursor -chloroauric acid-, the concentration of 0.01 M was fixed to evaluate a lesser value than the one used by Lechtman and other authors [3,6,7]. Moreover, in order to reach the optimum pH level of the electrolyte bath, we found that 0.13 M of sodium bicarbonate was enough to obtain a pH of 9.

Lechtman described the deposition temperature by associating it with a gentle boiling of the solution. However, to determine the exact temperature value in our experiments, previous studies that pointed out 80 °C and 90 °C for outstanding results were taken into consideration [8]. Here, 80 °C was taken as an optimal condition to avoid the faster evaporation of water at higher temperatures, which leads to nonuniform gilding processes. In this way, we were able to evaluate the amount of gold deposited in the copper coupon by XRF after 3 min and 6 min of reaction.

As shown in Figure 1, the plating process is not uniform in the whole sample due to the presence of several anodic regions on it (dark zones in the specimen). This phenomenon is caused by the oxidation of copper that rises because of the electrochemical replacement reaction. Copper behaves as an anode, transferring ions to the gold to be reduced at the solid-liquid interface. Equation (1) and Equation (2) show the half-reactions of copper and gold, respectively, and Equation (3) shows the balanced overall reaction in the cell. A similar replacement reaction was provided previously by Lechtman [3].

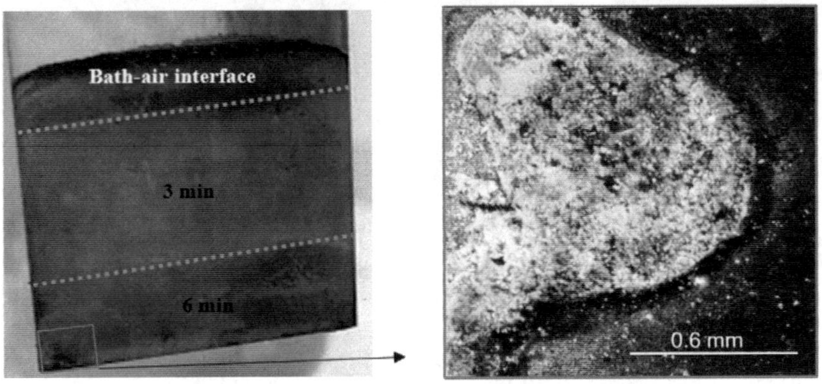

(a) (b)

Figure 1. (a) Image of the gilded copper sample after 3 min and 6 min of dipping time in the bath and (b) optical microscope image indicating the presence of an anodic region on the sample's surface.

$$Cu_{(s)} \rightarrow Cu^{2+}_{(aq)} + 2e^-, \tag{1}$$

$$Au^{3+}_{(aq)} + 3e^- \rightarrow Au_{(s)}, \tag{2}$$

$$2Au^{3+}_{(aq)} + 3\,Cu_{(s)} \rightarrow 2Au_{(s)} + 3Cu^{2+}_{(aq)}, \tag{3}$$

The plating of the copper sample was uniform and adherent throughout the immersed piece. On the other hand, at the bath-air interface, the water vapor released from the boiling solution generated an external anodic and nonuniform zone on the coupon (Figure 1). The electrochemical replacement reaction runs until the copper potential is uniform throughout the coupon. Thus, the anodic zones get covered at the maximum deposition time [9]. Figure 2 shows that after 6 min of deposition, the gold mass amount on the copper doubles its value regarding the 3 min test. Then, under our experimental conditions, the dipping time required to stop the electrochemical reactions goes beyond 6 min.

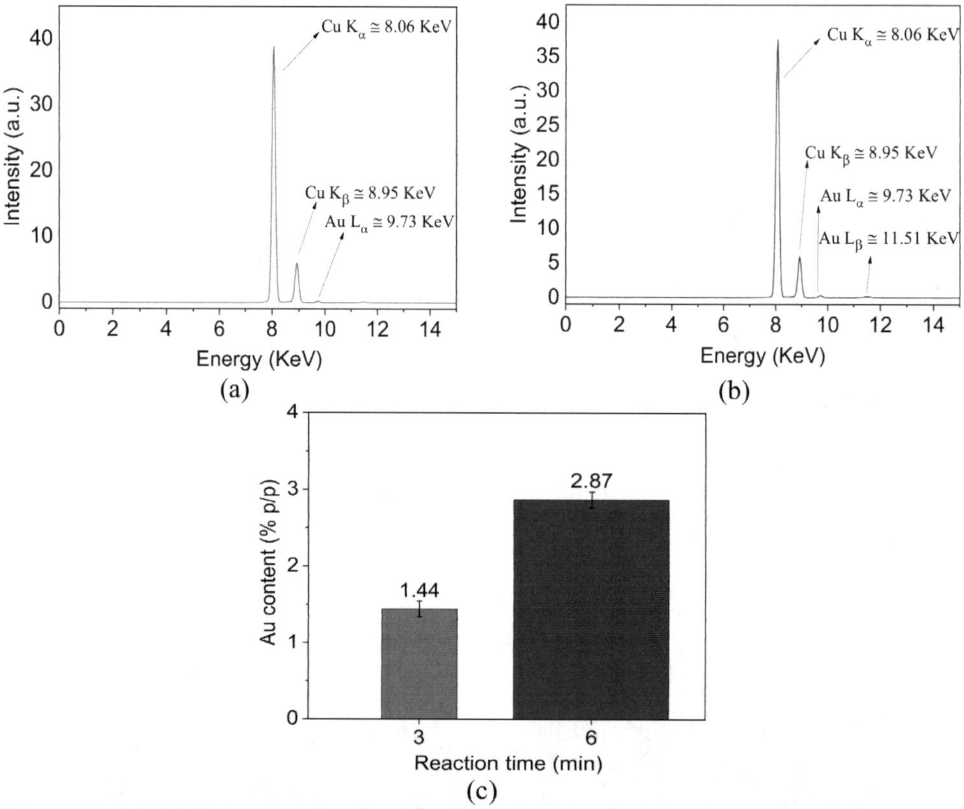

Figure 2. (a) XRF spectra after 3 min of reaction time, (b) XRF spectra after 6 min of reaction time, and (c) XRF mass percentage of gold deposited on copper for each testing time.

According to Lechtman, a good electrochemical replacement process should be characterized by tiny anodic zones surrounded by large cathodic areas. Those anodic regions are invisible to the human eye but big enough to sustain the electrons interchange between the substrate and the bath until they get completely covered [3]. She also believes that the presence of multiple pits on the copper surface could stimulate the conformation of the anodic regions. However, when searching for evidence in different excavated samples from Loma Negra, Piura valley, the current corroded state of the objects obstructed her scientific discussion.

The research work of Lechtman suggested that the electrochemical replacement process limits the thickness of the deposited gold to 2 μm and that temperatures from 500 °C to 800 °C are required to develop a diffusion zone at the copper interface to obtain a good adherence of the gold's film [3]. However, as shown in Figure 3, our results showed that it is feasible to obtain up to 7.5 μm of thickness for the deposited layer after 6 min in the electrolytic bath. The thicker plating layers developed in this work -even for 3 min of reaction (5.17 μm)- could be related to the introduction of higher temperatures in our experiments compared to Lechtman's protocol. In addition, the temperature of the process was enough to develop diffusion zones in the copper sample, almost doubling their length from 3 min to 6 min of reaction. This confirms the good adherence of the gilded layer obtained after the electrochemical replacement. It differs from Lechtman's argument even though further experimentation is lacking to evaluate this, and other issues associated with the thermal treatment, such as the grain size of the copper substrate.

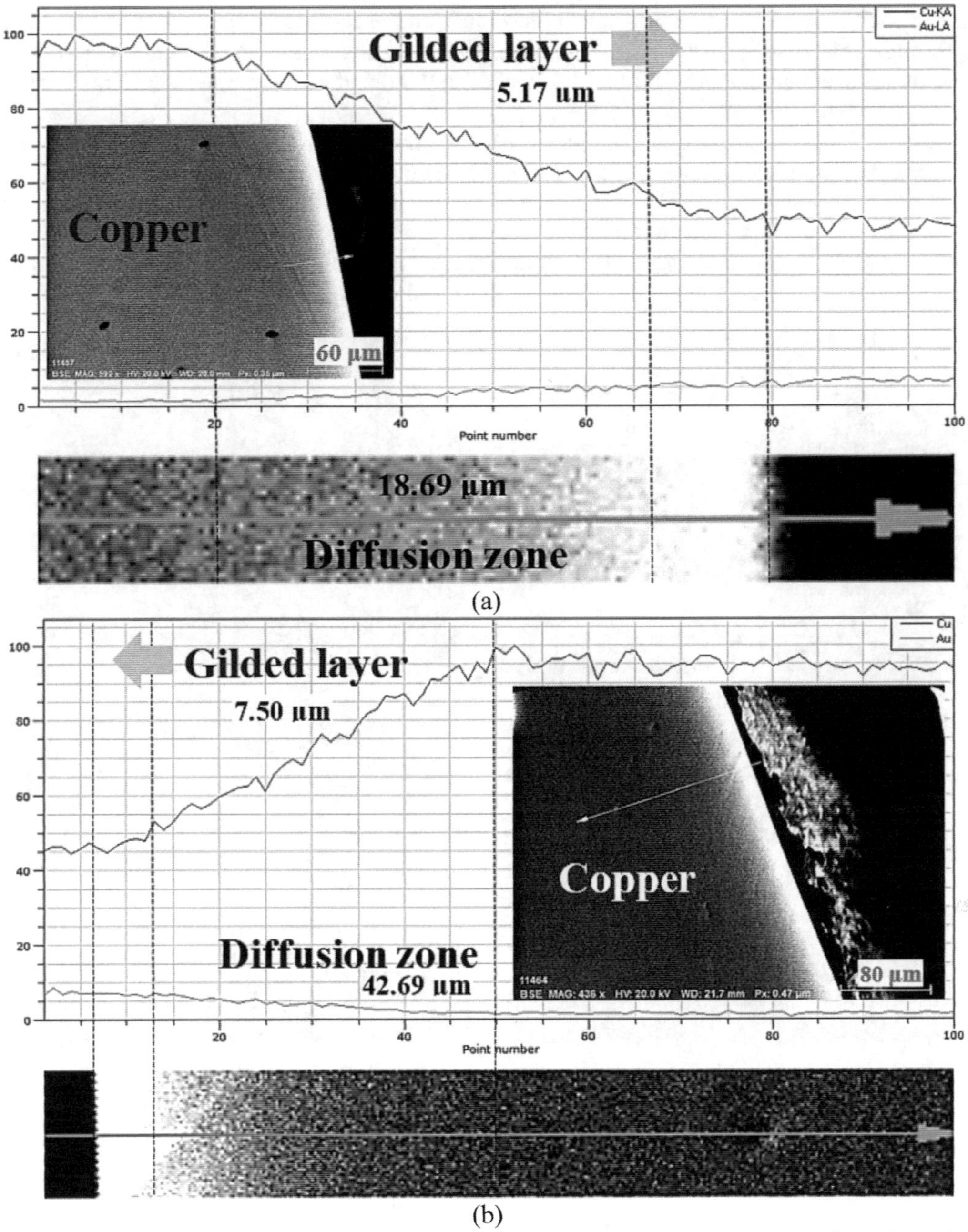

Figure 3. Backscattered images of the cross-section of the copper substrate after (a) 3 min and (b) 6 min in the electrolytic solution.

Lechtman's approaches to the techniques and methods used by Moche and Vicus goldsmiths are helpful to reconstruct their plating objects. However, the corrosion process that undergoes these pieces under environmental conditions could start with less difficulty when the thickness of gold is thinner. Besides, the corrosion processes could be accelerated by the humid and saline environments found on the northern coast of Peru [10,11]. As a result, some gilded objects would have been required to be re-plated with some frequency. Likewise, the presence of diverse corrosive agents such as chlorine in the burial environment enhances the diffusion of copper through gold, troublesome its conservation to the present [12]. As shown in Figure 4(a) and Figure 4(c), in burial environments, the solid diffusion of copper through gold is so important that it scratches the gold layer and erupts from it. Once the copper diffusion overpasses the cathodic protection, the degradation of the substrate accelerates, spreading transversely and underneath the layer of gold.

On the other hand, Figure 4(b) and Figure 4(d) show that the thickness of the gold layer reaches a value around 10 μm in Moches's gilded objects, thicker than those shown in Figure 3. Therefore, extended reaction times should be used to explore the possibility of higher thicknesses with the electrochemical replacement technique. Furthermore, we suggest that Moches and Vicus could have also used electroless or galvanic gilding techniques. But to sustain this statement, new experimental data associated with materials that these cultures could have used are missing.

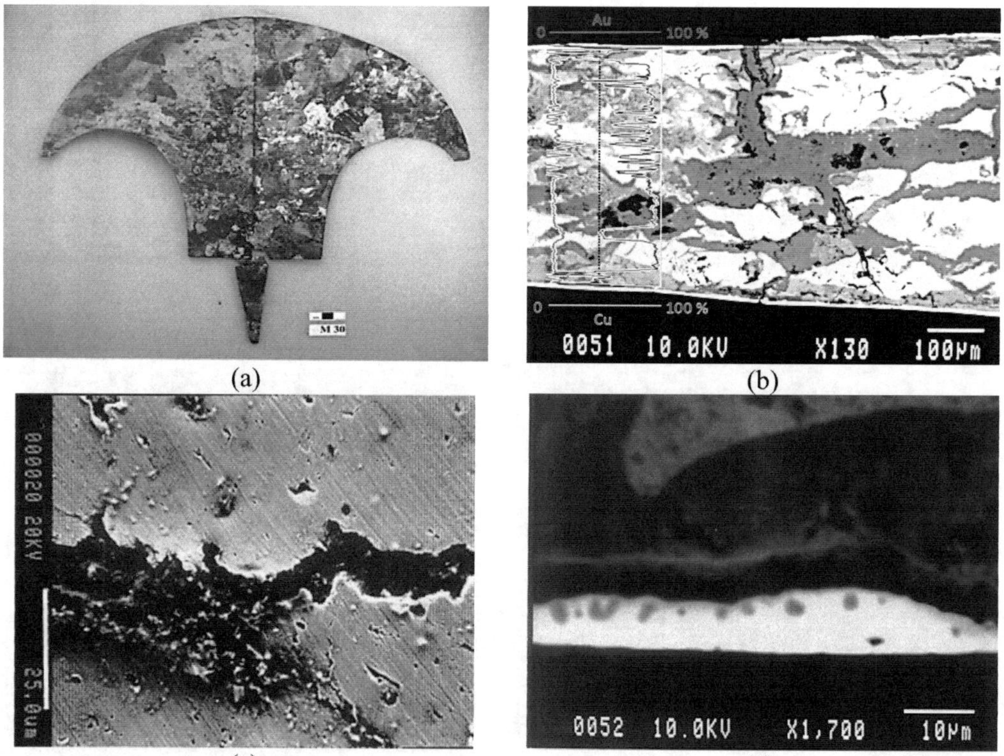

(a) (b) (c) (d)

Figure 4. (a) Moche style object excavated from Dos Cabezas archaeological site, (b) backscattered image of the cross-section, (c) secondary electron image from the surface of the sample, and (d) backscattered image of the gilding layer.

4. Conclusions

This research work provides scientific insights into the electrochemical replacement technique for plating copper objects with gold. We deposited adherent gold layers with thicknesses of 5.17 μm and 7.50 μm after dipping a copper sample for 3 min and 6 min into an electrolytic solution, respectively. However, several anodic regions were as well found in the copper substrate. This issue points out that the maximum deposition of gold on copper is beyond 6 min of reaction time. However, the thickness of gold observed in some archaeological samples exceeds the ones achieved in our trials. Futures studies should focus on extended reaction times as much as on assessing the corrosion rate of gilded coppers in burial and atmospheric environments. These arguments would reinforce the electrochemical replacement technique as the one used by Pre-Columbian goldsmiths. Also, plating techniques such as electroless and galvanic should be considered to enrich the scientific discussion. These research initiatives are essential for the development of conservation strategies to preserve our historical heritage.

Acknowledgments
The authors acknowledge to Concytec/Banco Mundial for their financial support in the Project "Mejoramiento y Ampliación de los Servicios del Sistema Nacional de Ciencia Tecnología e Innovación Tecnológica, 8682-PE" through its executing unit FONDECYT, contract number 035-2019. This project could not have been possible without the valuable collaboration of Juan Carlos Rodríguez Reyes, Ismat Shah, Carsten Benndorf, and Andrés De Leo, Centro de Investigación para la Conservación del Patrimonio, Universidad de Ingeniería y Tecnología (UTEC), Perú, whom we thank for their scientific advice. Also, Andrea Arana, Mario Carbajal, and Karinna Visurraga are thanked for their management support at UTEC.

References

[1] Bray W 1993 Techniques of gilding and surface-enrichment in pre-Hispanic American metallurgy *Metal Plating and Patination Cultural, Technical and Historical Developments* ed La Niece S, Craddock P (Boston: Butterworth-Heinemann) Chapter 16 p 182

[2] Watkinson D 2013 Conservation, corrosion science and evidence-based preservation strategies for metallic heritage artefacts *Corrosion and Conservation of Cultural Heritage Metallic Artefacts* ed Dillmann P, Watkinson D, Angelini E, Adriaens A (Cambridge: Woodhead Publishing) Chapter 2 p 9

[3] Lechtman H 1979 A pre-columbian technique for electronchemical replacement plating of gold and silveron copper objects *Journal of Metals* **31(12)** 154

[4] Lechtman H, Erlij A, Barry E J 1982 New perspectives on moche metallurgy: techniques of gilding copper at Loma Negra, Northern Peru *American Antiquity* **47(1)** 3

[5] Lechtman H 1984 Pre-Columbian surface metallurgy *Scientific American* **250(6)** 56

[6] Okinaka Y, Hoshino M 1998 Some recent topics in gold plating for electronics applications *Gold Bulletin* **31(1)** 3

[7] Wang Y, Cao X, Wang W, Mitsuzak N, Chen Z 2015 Immersion gold deposition from a chloroauric acid-choline chloride solution: deposition kinetics and coating performances *Surface and Coatings Technology* **265** 62

[8] Schwarz A, *et al.* 2010 Gold coated para-aramid yarns through electroless deposition *Surface and Coatings Technology* **204(9)** 1412

[9] Djokić S S, Cavallotti P L 2010 Electroless deposition: theory and applications *Modern Aspects of Electrochemistry* ed Vayenas C G, White R E, Gamboa-Aldeco M E (New York: Springer) Chapter 6 p 251

[10] Alviz-Meza A, Rodríguez-Reyes J C F, Chacaliaza J A 2021 Thermodynamic stability of gilded copper and pigments at high relative humidity and at environmental conditions of Lima, Perú *Journal of Physics: Conference Series* **1938(1)** 12003:1

[11] Alviz-Meza A, Kafarov V, Peña-Ballesteros D Y 2017 Study of the continuous corrosion in an oxidation environment derived from the theoretical combustion products in a refinery. Case study: ferritic steel ASTM A335 P91 *Journal of Physics: Conference Series* **935(1)** 012057:1

[12] Pusic S P 1993 Diffusion of copper into gold *IEEE Instrumentation and Measurement Technology Conference* (Irvine: IEEE) p 114

Numerical study of the physical processes of gas leakage in the compression ring in diesel engines

J A Pabón León[1], J P Rojas Suárez[1], and M S Orjuela Abril[1]
[1] Universidad Francisco de Paula Santander, San José de Cúcuta, Colombia

E-mail: sofiaorjuela@ufps.edu.co

Abstract. In this research, the construction of a numerical model is proposed for the analysis of the friction processes and the thickness of the lubrication film present in the compression ring of internal combustion engines. The model is built using MATLAB software, and three load conditions are used as reference (2 Nm, 4 Nm, and 6 Nm) with a rotation speed of 3600 rpm, which correspond to a stationary single-cylinder diesel engine. Comparison between model estimates and experimental results show that the development model could predict the actual engine conditions. The deviation between the numerical model and the experimental data was 17%. It was shown that the increase in engine load causes a 16% increase in the friction force of the compression ring, which implies a 50% increase in power loss due to friction processes. In general, the model developed allows the analysis of the friction processes in the compression ring and its effect on the lubrication film, considering the leakage of the combustion gases. In this way, the construction of a more complex mathematical model is achieved, which allows improving the precision in the analyzes related to the interaction between the compression ring and the cylinder liner.

1. Introduction

The function of the compression ring is to establish a tight seal between the piston skirt and the cylinder liner; in this way, the leakage of combustion gases is avoided since this type of leakage leads to pressure losses inside the combustion chamber and a reduction in the energy efficiency of the engine. However, high interaction between the compression ring and the cylinder liner causes an increase in friction forces, and therefore a loss in engine power [1,2]. Due to the above, it is necessary to establish an adequate interaction between the cylinder liner and the compression ring. Studies available in the literature show that approximately 11.5% of fuel energy is used to overcome engine friction [3]. It is estimated that the greatest mechanical losses are located in the piston assembly, which represent between 38% – 68% of the total mechanical losses [4]. The piston rings are responsible for 45% of the losses in the piston assembly. Therefore, it can cause a significant deterioration in the performance of the engine [5].

Due to the influence that the interaction between the compression ring and the cylinder liner has on the performance of the engine, it is necessary to use models that allow predicting the energy losses experienced by the engine [6]. In the literature, different formulations with different degrees of complexity are shown to study the behavior of the compression ring; Baelden and Tian [7] used finite element analysis to evaluate the influence of ring deformation on lubricant transport. Baker, *et al.* [8] developed a three-dimensional elasto-dynamic ring model; From the results obtained, it was shown that the dynamics of the piston rings affect the wear of the piston liner, the degradation and consumption of oil, and the formation of hydrocarbon products. Xu, *et al.* [9] indicated that friction processes have direct implications on the surface wear of the cylinder liner. Peng, *et al.* [10] reported that wear and friction between the compression ring and the cylinder liner could cause engine instability. Baker, *et al.* [11] studied the effect of compression of the piston ring on the pressure loss of the combustion chamber. Tian [12] constructed a finite element model to evaluate oil transport and piston ring lubrication. The results indicate that the dynamics of the rings significantly influence the tribological characteristics present in the contact of the cylinder liner and the piston skirt. Additionally, it was concluded that the study of the dynamics of the compression ring is a requirement to accurately understand the power losses of the engine.

Combustion gas leakage is a phenomenon present between the compression ring and the cylinder liner, which affects the dynamic behavior of the ring. This phenomenon causes energy losses, an increase in the temperature of the engine body, and an increase in the oil temperature. Additionally, it causes lubrication disturbances, increased oil consumption, accelerated wear of oil and components, and increased emissions, especially particulate matter and hydrocarbons [5]. Due to the aforementioned, the analysis of the dynamics of combustion gas leaks is essential for a more precise compression of the energy losses present during the combustion process of internal combustion engines.

According to the above, the present investigation proposes the development of a mathematical model to describe the dynamic and tribological characteristics present between the compression ring and the piston liner, considering combustion gas leaks, which are generally neglected in this type of analysis. The foregoing, with the aim of providing physics with a model that allows better precision and understanding of the energy losses caused by the compression ring of the piston.

2. Methodology

This section describes the numerical models used in the study, which involve hydrodynamic lubrication and combustion gas leak models; additionally, the numerical procedure used in the investigation is specified.

2.1. Hydrodynamic lubrication model

The hydrodynamic behavior between the compression ring and the cylinder liner was modeled by using the Reynolds equation, as shown in Equation (1) [13].

$$\frac{\partial}{\partial x}\left(\frac{h^3}{12\mu}\cdot\frac{\partial P}{\partial x}\right)+\frac{\partial}{\partial y}\left(\frac{h^3}{12\mu}\cdot\frac{\partial P}{\partial y}\right)=v_p\cdot\frac{\partial h}{\partial x}+\frac{\partial h}{\partial t}, \tag{1}$$

where P is the pressure distribution, h is the film thickness, μ is the kinematic viscosity and v_p is the sliding velocity; the dynamic viscosity (η) and the density (ρ) of the lubricant are determined from the correlations proposed by Houpert [14] and Dowson–Higginson [15], which are indicated in Equation (2) and Equation (3).

$$\eta=\eta_o\cdot e^{\left[\ln\left(\frac{\eta_o}{\eta_\infty}\right)\times\left(\left\{\frac{T-138}{T_o-138}\right\}^{-s_o}\left\{1+\frac{P-P_o}{c_p}\right\}^{Z}-1\right)\right]}, \tag{2}$$

$$\rho=\rho_o\cdot(1-\gamma[T-T_o])\cdot\left(1+\frac{6\times10^{-10}\cdot(P-P_o)}{1+1.7\times10^{-9}\cdot(P-P_o)}\right), \tag{3}$$

where $c_p=1.98\times10^{8}$ Pa and $\eta_\infty=6.31\times10^{-5}$ Pa $\cdot$ s are constants of the model, s_o is the thermo-viscosity index, Z is the piezo-viscosity index, and γ is the coefficient of thermal expansion of the lubricant. The subscript o indicates atmospheric conditions. The sliding velocity (v_p) of the compression ring is considered equal to the velocity of the piston. Therefore, v_p can be determined by Equation (4) [16].

$$v_p=r\cdot\omega\cdot\sin\theta+\frac{r\cdot\omega\cdot\sin\theta\cdot\cos\theta}{\sqrt{\frac{1}{r}-\sin^2\theta}}, \tag{4}$$

where r is the crank radius, ω is the crankshaft angular velocity, l is the connecting rod length, and θ is the crank angle. The power loss (P_l) in the compression ring is determined by Equation (5) [17].

$$P_l = (f_v + f_b) \cdot |v_p|, \tag{5}$$

where f_v is the viscous friction and f_b is the ring boundary friction, respectively.

2.2. Combustion gas leak model

As the pressure inside the combustion chamber increases, the combustion gases tend to be directed into the grooves between the compression ring and the piston. The mass flow rate ($\dot{m}_g$) is determined considering an isentropic behavior, as shown in Equation (6) [18].

$$\dot{m}_g = f_m \frac{A_g \cdot C_f \cdot P_u}{\sqrt{R \cdot T_u}}, \tag{6}$$

where C_f is the coefficient of discharge, A_g is the downstream ring end gap area, f_m is the compressibility factor, T_u is the orifice upstream temperature, R is the ideal gas constant and P_u is the gas pressure upstream. The coefficient of discharge (C_f) and the compressibility factor (f_m) are calculated by Equation (7) and Equation (8) [18].

$$C_f = 0.85 - 0.25\left(\frac{P_d}{P_u}\right)^2, \tag{7}$$

$$f_m = \left\{ \begin{matrix} \sqrt{\gamma} \cdot \left(\frac{2}{\gamma+1}\right)^{\frac{\gamma+1}{2(\gamma-1)}}, & \frac{P_d}{P_u} > \left(\frac{2}{\gamma+1}\right)^{\frac{\gamma}{(\gamma-1)}} \\ 0.85 - 0.25\left(\frac{P_d}{P_u}\right)^2, & \frac{P_d}{P_u} > \left(\frac{2}{\gamma+1}\right)^{\frac{\gamma}{(\gamma-1)}} \end{matrix} \right\}, \tag{8}$$

where P_d is the gas pressure downstream, and γ is the ratio of the specific heats; the power loss associated with the leak of combustion gases (P_k) is determined by Equation (9) [19].

$$P_k = \frac{\dot{m}_g \cdot R \cdot T_u \cdot \gamma}{\gamma - 1} \cdot \left(1 - \left[\frac{P_d}{P_u}\right]^{\frac{\gamma-1}{\gamma}}\right). \tag{9}$$

2.3. Numerical procedure

The equations described in section 2.1 and section 2.2, are solved using the ode45 solver of the MATLAB software; Figure 1 describes the flow diagram of the numerical model.

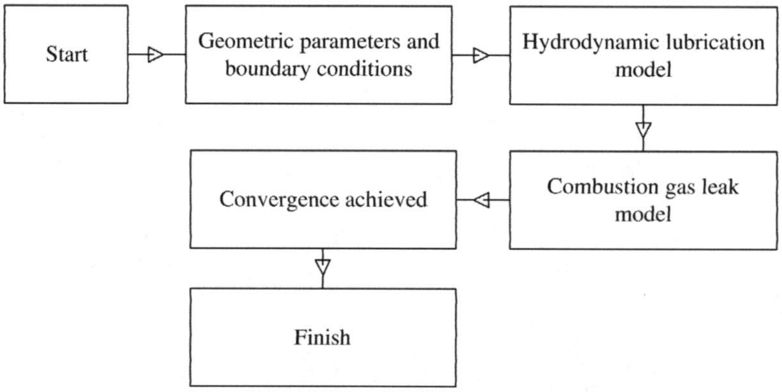

Figure 1. Numerical model flow diagram.

The reference conditions of the numerical model are based on the technical and operational characteristics of a stationary single-cylinder diesel engine. The properties of the engine are described in Table 1. A constant rotational speed of 3600 rpm and three load conditions (2 Nm, 4 Nm, and 6 Nm) were selected for the numerical analysis of the engine.

Table 1. Technical specifications of the engine.

Characteristics	Value
Model	SK – MDF 300
Manufacturer	SOKAN
Stroke	63 mm
Bore	78 mm
Intake system	Naturally aspirated
Displaced	300 cc
Injection system	Direct injection

3. Results

This section describes the results and discussions obtained for the present investigation; the analysis carried out includes the evaluation of the friction force and the analysis of the loss power associated with this force. Additionally, a study of the change in the thickness of the lubrication film is carried out, as well as the pressure conditions present in the compression ring.

3.1. Experimental validation

To guarantee the reliability of the model, a comparison was made with the friction force during the combustion cycle measured experimentally. This measurement was carried out with a rotation speed of 3600 rpm and a load of 6 Nm. The results of the comparison between the model and experimental data are shown in Figure 2.

From Figure 2, it was observed that the model could represent the real behavior experienced by the compression ring of the engine. The adequate similarity was evidenced in both magnitude and trend between the model and the experimental data. Analysis of the mean relative error shows a deviation of 17%. Due to the difficulty for the actual measurement of the friction force on the piston, the error reached is considered adequate [20].

The analysis of the pressure inside the combustion chamber throughout the combustion cycle is indicated in Figure 3 for the load conditions of 2 Nm, 4 Nm, and 6 Nm. The results indicate that pressure tends to increase with increasing engine load levels. In general, it was observed that a 2 Nm increase in engine load causes an 11% increase in peak combustion pressure. This may imply greater ease of combustion gas leakage and greater power losses due to friction processes between the compression ring and the cylinder liner.

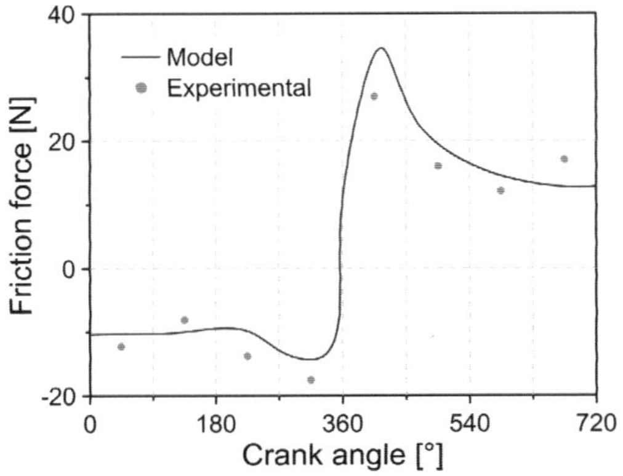

Figure 2. Comparison between the friction force obtained experimental vs. model.

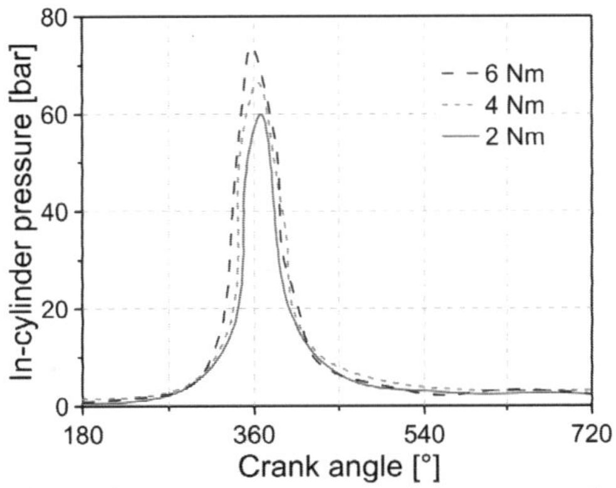

Figure 3. Average pressure of the pneumatic network.

Figure 4 shows the influence of the engine load on the friction force present in the compression ring; it was observed that the maximum magnitude of the friction force occurs during the expansion process (360° – 540°) for all load conditions. This can be attributed to the reduction in the lubrication film and the high acceleration experienced by the piston during this stage. In general, the increase in engine load causes an increase in the friction force. For the load conditions of 2 Nm, 4 Nm, and 6 Nm, a maximum force of 25.6%, 29.4%, and 34.8% was recorded, respectively.

The friction force causes a loss of power, which can be seen in Figure 5. Increasing the engine load tends to increase the loss of engine power. This is directly associated with the greater magnitude of the friction force, as shown in Figure 4. In general, the greatest losses occur during the middle of each stage of combustion. This is due to the mixed lubrication regime condition present during these parts of the cycle.

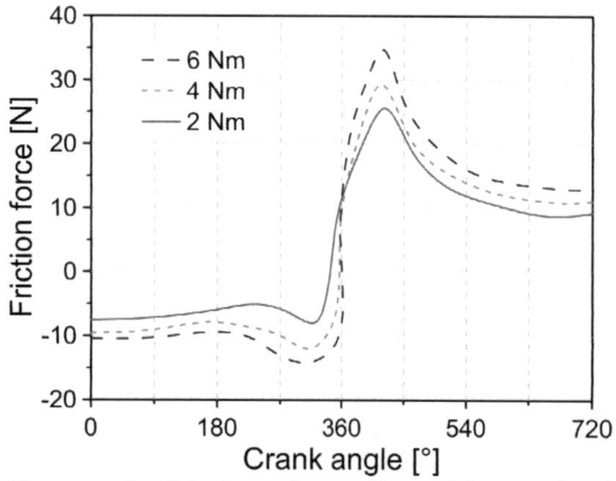

Figure 4. Friction force for different load conditions.

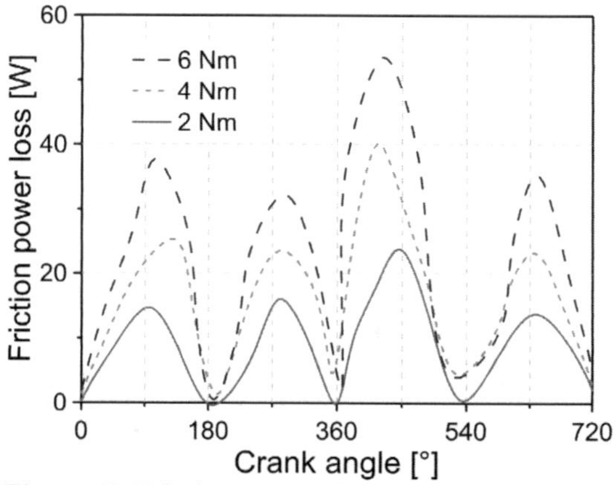

Figure 5. Friction power loss for different load conditions.

Figure 6 indicates the minimum lubrication film thickness during the combustion cycle for different load conditions. In general, the smallest thickness of the lubrication film was located at the angles of 180°, 360°, 540°, and 720°, which coincide with the change in the direction of movement of the piston. This facilitates the formation of a mixed lubrication regime, and therefore, the reduction in the minimum thickness of the lubrication film. A minimum lubrication film thickness of 1.50 μm, 2.19 μm, and 3.23 μm was reported for a load of 2 Nm, 4 Nm, and 6 Nm, respectively.

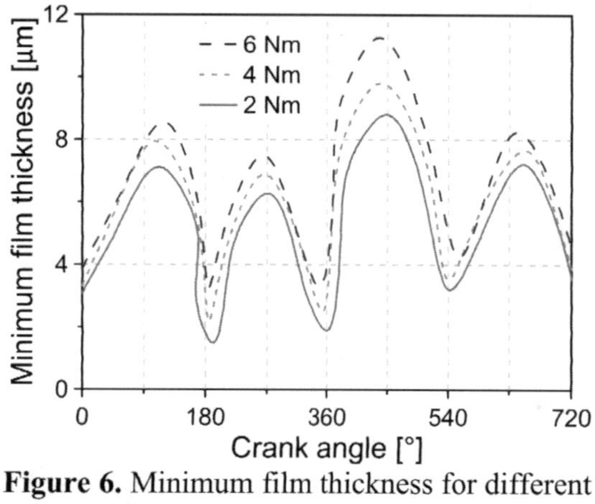

Figure 6. Minimum film thickness for different load conditions.

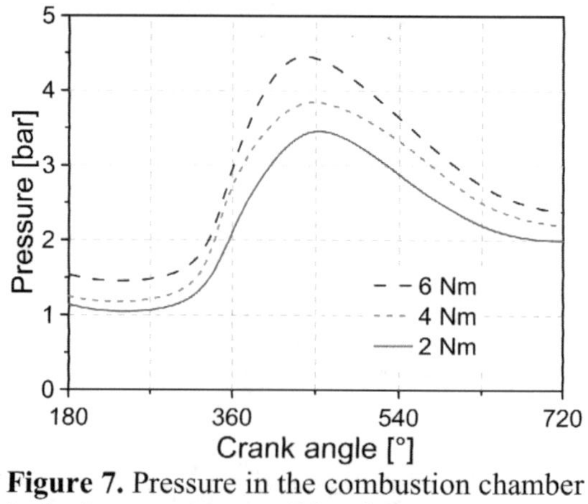

Figure 7. Pressure in the combustion chamber in the bottom groove of the ring.

Figure 7 indicates the localized pressure in the bottom groove between the compression ring and the piston during the combustion cycle. From the results obtained, it is shown that the pressure curves present a similar behavior for the different load conditions. The maximum pressure peak was located approximately at an angle of 440°, which corresponds to the descent of the piston during the expansion stage; for a load of 2 Nm, 4 Nm, and 6 Nm, maximum pressure of 3.45 bar, 3.83 bar, and 4.44 bar was obtained, respectively. This pressure increase implies a greater amount of flow gas leakage.

4. Conclusions

In this research, the development of a numerical model for the analysis of the friction processes experienced by the compression ring is proposed, considering the hydrodynamic lubrication characteristics and the combustion gas leaks present in the combustion chamber jointly. The development of the model was carried out using MATLAB software, taking as reference three load conditions (2Nm, 4 Nm, and 6 Nm) with a rotation speed of 3600 rpm.

Comparison between model estimates and experimental results show that the development model could predict the actual engine conditions. In general, the deviation between the numerical model and the experimental data was 17%, which is acceptable for the type of study carried out. The model allowed the analysis of friction forces and the associated energy loss during the combustion cycle. The increase in engine load was shown to cause a 16% increase in the friction force of the compression ring. The above implied an increase of approximately 50% in the power loss due to friction processes. The model allowed to analyze the minimum thickness of the lubrication film, showing that the change in the direction of the piston movement causes the greatest risk of wear in the engine. For the load conditions of 2 Nm, 4 Nm, and 6 Nm, a minimum lubrication thickness of 1.50 μm, 2.19 μm, and 3.23 μm, was reported, respectively. Additionally, it was observed that the increase in engine load causes a 15% increase in the pressure of the bottom ring groove.

In general, the model developed allows the analysis of the friction processes in the compression ring and its effect on the lubrication film, considering the leakage of the combustion gases, which is usually ignored in this type of analysis. In this way, the construction of a more complex mathematical model is achieved, which allows improving the precision in the analyzes related to the interaction between the compression ring and the cylinder liner.

References

[1] Consuegra F, Bula A, Guillín W, Sánchez J, Duarte Forero J 2019 Instantaneous in-cylinder volume considering deformation and clearance due to lubricating film in reciprocating internal combustion engines *Energies* **12(8)** 1437

[2] Hernández B, Maestre D, Pardo C, Fonseca M D S, Pabón J 2021 Influence of compression rings on the dynamic characteristics and sealing capacity of the combustion chamber in diesel engines *Lubricants* **9(3)** 25

[3] Holmberg K, Andersson P, Erdemir A 2012 Global energy consumption due to friction in passenger cars *Tribology International* **47** 221

[4] Tormos B, Martin J, Pla B, Jimenez-Reyes A J 2020 A methodology to estimate mechanical losses and its distribution during a real driving cycle *Tribology International* **145** 106208

[5] Koszalka G, Hunicz J 2021 Comparative study of energy losses related to the ring pack operation in homogeneous charge compression ignition and spark ignition combustion *Energy* **235** 121388

[6] Forero J D, Ochoa G V, Alvarado W P 2020 Study of the piston secondary movement on the tribological performance of a single cylinder low-displacement diesel engine *Lubricants* **8(11)** 97

[7] Baelden C, Tian T 2014 A dual grid curved beam finite element model of piston rings for improved contact capabilities *SAE International Journal of Engines* **7(1)** 156

[8] Baker C, Rahmani R, Theodossiades S, Rahnejat H, Fitzsimons B 2015 On the effect of transient in-plane dynamics of the compression ring upon its tribological performance *Journal of Engineering for Gas Turbines and Power* **137(3)** 032512

[9] Xu Y, Zheng Q, Geng J, Dong Y, Tian M, Yao L, Dearn K D 2019 Synergistic effects of electroless piston ring coatings and nano-additives in oil on the friction and wear of a piston ring/cylinder liner pair *Wear* **422** 201

[10] Peng Y, Xu Y, Geng J, Dearn K D, Hu X 2017 Tribological assessment of coated piston ring-cylinder liner contacts under bio-oil lubricated conditions *Tribology International* **107** 283

[11] Baker C, Rahmani R, Karagiannis I, Theodossiades S, Rahnejat H, Frendt A 2014 Effect of compression ring elastodynamics behaviour upon blowby and power loss *SAE Technical Paper* **2014-01-1669** 1

[12] Tian T 2002 Dynamic behaviours of piston rings and their practical impact. Part 2: Oil transport, friction and wear of ring/liner interface and the effects of piston and ring dynamics *Proceedings of the Institution of Mechanical Engineers, Part J: Journal of Engineering Tribology* **216(4)** 229

[13] Patir N, Cheng H S 1978 An average flow model for determining effects of three-dimensional roughness on partial hydrodynamic lubrication *Journal of Lubrication Technology* **100(1)** 12

[14] Houpert L 1985 New results of traction force calculations in elastohydrodynamic contacts *Journal of Tribology* **107(2)** 241

[15] Dowson D, Higginson G R 1959 A numerical solution to the elasto-hydrodynamic problem *Journal of Mechanical Engineering Science* **1(1)** 6

[16] Rahnejat H 2005 *Multi-Body Dynamics. Vehicles, Machines, and Mechanisms* (Bury St Edmunds: John Wiley & Sons Ltd)

[17] Zavos A, Nikolakopoulos P G 2021 Investigation of the top compression ring power loss and energy consumption for different engine conditions *Tribology-Materials, Surfaces & Interfaces* 1

[18] Kurbet S N, Kumar R K 2007 Finite element modelling of piston-ring dynamics and blow-by estimation in a four-cylinder diesel engine *Proceedings of the Institution of Mechanical Engineers, Part D: Journal of Automobile Engineering* **221(11)** 1405

[19] Woods W A 1991 On the role of the harmonic mean isentropic exponent in the analysis of the closed-cycle gas turbine *Proceedings of the Institution of Mechanical Engineers, Part A: Journal of Power and Energy* **205(4)** 287

[20] Gore M, Theaker M, Howell-Smith S, Rahnejat H, King P D 2014 Direct measurement of piston friction of internal-combustion engines using the floating-liner principle *Proceedings of the Institution of Mechanical Engineers, Part D: Journal of Automobile Engineering* **228(3)** 344

Journal of Physics: Conference Series **2118** (2021) 012017 doi:10.1088/1742-6596/2118/1/012017

Using implementation of artificial intelligence in estimating the exchange rate in the foreign exchange market

D F Devia Narvaez[1], **R Ospina**[2], **and F Mesa**[1]
[1] Universidad Tecnológica de Pereira, Pereira, Colombia
[2] Universidad Industrial de Santander, Bucaramanga, Colombia

E-mail: femesa@utp.edu.co

Abstract. The Foreign Exchange currency market is considered the most liquid market in the world due to the amount of money that is traded every day. Researchers and investors are interested in having a system that allows predicting the direction of the exchange rate of a pair of currencies in order to create a strategy that allows them to profit from the market. Since the movement of the exchange rate between currencies generates a high frequency time series, recurrent neural networks have gained strength in the prediction of these markets, since they have the characteristic of generating models of non-linear systems. In this document, the study of the prediction of the direction of the exchange rate of different currencies was carried out using the recurrent auto-regressive neural network with exogenous inputs, which has the characteristics to obtain a predictive model of a time series. The tests were carried out in the Euro / Dollar pair in the time frame of one hour, demonstrating the potential of this neural network architecture. These techniques are products of the biological and physical sciences in industrial processes.

1. Introduction

The foreign exchange (Forex) market is a decentralized global market use to trade between exchange pairs. The key participants are the international banks, hedge funds, commercial companies, central banks, and of course, minimum exchange traders and investors [1,2]; as a result, the exchange rates between currencies are constantly changing, shaping a market of high frequency and volatility that has many directional changes. Market participants enter orders to profit from bullish or bearish movements, depending on the operation case as buying or selling. For this reason, have a way to predict the market price direction between currencies is very important to reduce risks and maximize the participants' profits.

Many approaches have been used to profit from the market, such as technical analysis, fundamental analysis, statistical analysis, and predictive calculation techniques, each of them has their influence on optimization, market analysis, political science, etc. econometrics all of them permeated by the techniques of physics and mathematics.

One theory affirms that the market movement can be analyzed as a chaotic nonlinear time series that seems random but can be predictable; another opposite theory, the efficient market theory, claims that the market is random and unpredictable. Many studies supported the first theory and had applied artificial intelligence techniques like the neural network that have shown promising results when predicting series of times with these characteristics. In this work, recurrent neural networks are used based on diverse authors [3,4]. These networks have the capabilities to predict future values of nonlinear time series. The presence of recurrent feedback at the neural network seems a positive factor in the financial time series forecasts [5]. This paper focused on applying recurrent neural networks to predict the pair Euro/Dollar (EUR/USD) and for different architectures of the neuronal recurrent auto regressive (NARX) network.

2. Recurrent neural networks usage at financial markets

Regarding its specific use in the world in recent years, Zheng [6] mentioned one of them, who designed the called neural network of Elman, a typical regression network appliable to process sequential data. The created network was used to predict the action price or its fluctuance. At [7], the authors applied neural network models of multilayer perceptron to predict the action prices from the Information Technology Indian companies; at [5], the studied problem was the prognostic and increase of prices. As data analyzed were taken indexes like Standard and Poor´s 500 (S & P 500), Nasdaq100, and Dow Jones. Different multilayer perceptron setups were used, with different numbers of hidden neurons. The results demonstrated the possibility of forecasting the sign of the price increments with a slightly higher probability of 50%; at [8], the problem studied was to predict the price incremental of S & P 500. Historical data was taken from index S & P 500 for the period 1928 to 1993. Many neural networks have been built and trained with different setups. As a result, surprisingly, almost all the neural networks made a profitable prediction; at [9], the sign of the prices increase was forecasted. Germany stock index (DAX) and financial times stock exchange (FTSE100) indexes were used for some time. The Elman recurrent neural network was chosen. Finally, the authors concluded that the neural networks cannot give better results than simple models, like the Markov model; at [10], a neural network forecast was made of currencies exchanges in Forex. The analyzed data was selected from weekly data of Australian dollar (AUD), Swiss franc (CHF), German mark (DEM), Pound sterling (GPB), and Japanese yen (JPY). Two neural networks were used in this study: the first model is the multilayer perceptron, its inflows fed exchange rate values with some delay, and the second model was the multilayer perceptron, with inflows from which exchange rate of mobile media values were fed with time windows (from 5 to 120 tics). Statistical estimates of the quality of the prognosis were made, showing, for the last model, that neural networks can give good results.

3. Currencies market

In the foreign exchange market (forex: foreign exchange) when a participant enters a purchase order, it is known as a unity purchase from the primary currency and selling the second currency. Depending on the purchase or sale operation, the market participants enter orders to profit from the bearish and bullish movements. The actual price movement is called ticks that show the existence of transactions between sellers and buyers. There can be numerous ticks per second during active markets, and in slow markets, there can be minutes without a tick. The market formation and the participation option are made through Forex brokers. Through these platforms, the investors see the market and execute their sell or buy orders. In Figure 1 the Forex signal is shown, which is an economic time series that is made up of two composite structures: a long-term trend and a short-term high-frequency oscillation.

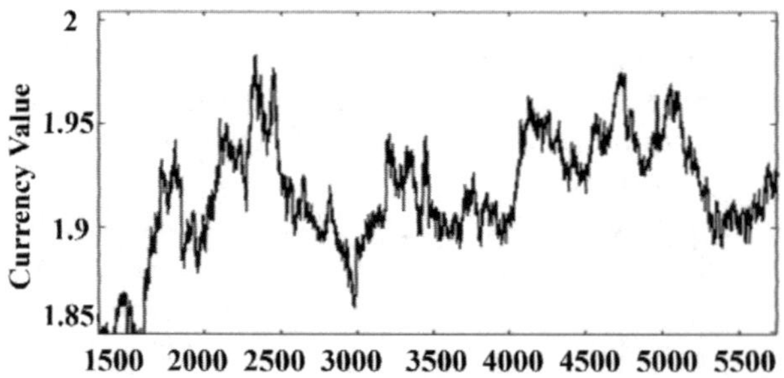

Figure 1. Forex market time series.

4. Recurrent neural networks

Recurrent neural networks [11] are, in principle, a computational model more potent than the classic neural networks; neural, take an advanced network architecture and add feedback connections to the previous layers. Recurrent neural networks are helpful in situations in which a temporal relation (time-dependent) exists in the data; Jordan (1986) and Elman (1990) introduced this architecture type. Neural networks are trained by a standard counter-propagation algorithm, except that the patterns are always presented in sequential order in time. In this way, the network sees the previous knowledge that has over the last posts. Furthermore, the presence of recurrent feedback at the neural network seems a positive factor in the prognostic of financial time series [12]. Thus, the recurrent neural network has a "deeper memory" than an advanced neural network.

These networks have short- and long-term memory blocks depending on the architecture complexity. A nonlinear autoregressive with exogenous input (NARX) is used in this work; it is a frequent dynamic network used commonly to model time series. There are many applications for the NARX network. Furthermore, it can be used to predict the next value of the input signal. Also, the output destiny is a version without noise of the input signal, used as a nonlinear filter. The NARX networks can be applied to model nonlinear dynamic systems. Figure 2 shows a result diagram of the network, using an advanced bilayer network for the approach; this implementation also allows a vectorial ARX, where the input and the output can be multi-dimensional.

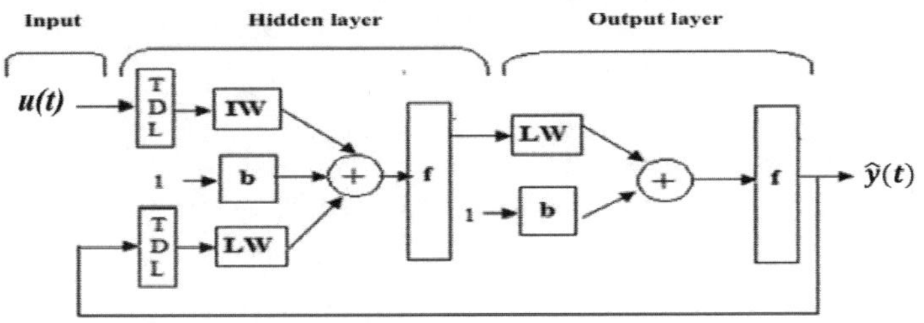

Figure 2. The NARX architecture [12].

The NARX network can be used in parallel architecture (closed-loop) and series-parallel architecture (open-loop), as shown in Figure 3. In this work, the open loop was used to process the training, allowing the network to be trained with the standard propagation algorithm applied to the multilayer perceptron. For the network training of the open loop, the data series u is presented like n data of the training Equation (1) [13].

$$u = f(u(t-1), u(t-2), \ldots, u(t-n)), \tag{1}$$

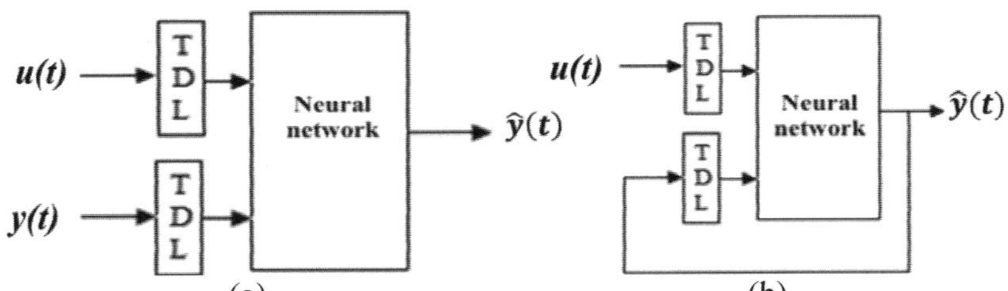

(a) (b)

Figure 3. Networks architectures (a) open-loop, and (b) close-loop [12].

The output of a supervised train in a neural network should equal the series u(t) values; in the case of the close loop, the estimation is giving by the data series y, where the training target is to search that the future known data u(t) equals the estimated data by the neural network y(t). Therefore, in Equation (2) [13], the open loop is fed by the input too with the data y(t) = u(t). When the trained network, the loop is closed to predict m future data of u series.

$$y = f(y(t), y(t+1), y(t+2), \ldots, y(t+m)). \tag{2}$$

5. Proposed methodology for the estimation

A nonlinear autoregressive with exogenous input (NARX) from MATLAB© is proposed for this work to predict the direction and the future value of the currency exchange rate.

5.1. Network training

Figure 4 shows the order process scheme for the training phase.

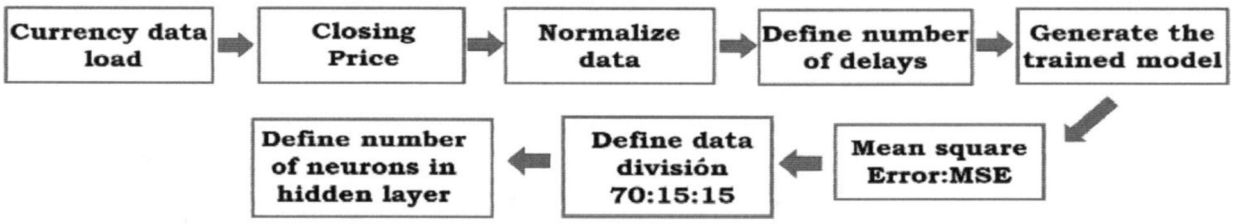

Figure 4. Training phase of the structure.

5.2. Normalize data

The data are normalized in the range [-1,1] Equation (3) [13].

$$U_n(t) = \frac{u(t) - \text{mean}(u)}{\max(u) - \min(u)}, \tag{3}$$

5.3. Number of delays and number of neurons

The network's prediction performance was evaluated in this work for different delay values and the number of neurons.

5.4. Data division

The data from the series is divided into the following proportions 70:15:15, which are the training data, validation data, and test data; with these proportions, the neural network has a data series to validate and probe the model.

5.5. Network validation

The mean square error is used to validate the prediction's quality. Equation (4) [13], shows the propagation algorithm backward as a criterion to update the networks' weights.

$$\text{MSE} = \frac{1}{2}\sum(y(t) - u(t))^2, \tag{4}$$

5.6. Making the prediction

Figure 5 shows the ordered process scheme for each prediction stage.

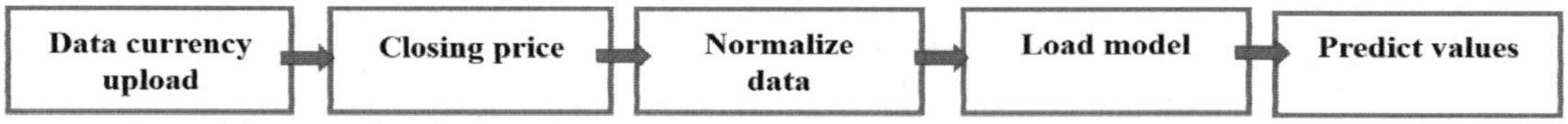

Figure 5. Architecture for the prediction stage.

6. Implementation and results

The prediction is tested for the pair EUR/USD. The closing price is used as data series for the temporality of one-hour H1. With the data from (31/05/2017) to (24/01/2019), for a total data amount of 10036. Following the structure mentioned in Figure 5, the first to do is normalize the data using Equation (3) in the range [-1,1] as shown in Figure 6.

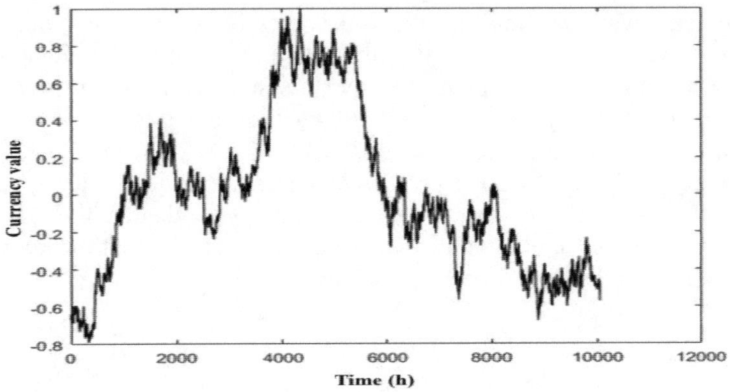

Figure 6. EUR/USD historical, H1 temporality normalized.

Table 1 registers the training tests and prediction varying for different delay values and number of neurons from the hidden layer. As test measurements, three indexes are calculated.

a) The mean square error (MSE), obtained at the training stage.
b) The mean square error to predict one forward data value.
c) The mean square error to predict five forward data values.

Table 1. Test results.

Test	Delays	N° Neurons C. O.	MSE	Mistake one step	Mistakes five steps
1	3	5	0.00018334	0.09523	1.51555
2	3	10	0.00018328	0.09507	1.51170
3	3	15	0.00018311	0.09530	1.50772
4	5	5	0.00018326	0.09581	1.51639
5	5	10	0.00018342	0.09485	1.51097
6	5	15	0.00018314	0.09574	1.51662
7	10	5	0.00018292	0.09467	1.50958
8	10	10	0.00018290	0.09554	1.50929
9	10	15	0.00018390	0.09548	1.50539

The set of n data estimated in the forward direction through the closed-loop prediction of the neural network is calculated with Equation (5) [13].

$$\hat{Y}_t \rightleftarrows= \{y(t+1), y(t+2), \ldots, y(t+n\}. \tag{5}$$

The prediction's means square error is Equation (6) [13].

$$MSE_n = \frac{1}{2}\sum_{t=\rightleftarrows 1}^{m}\sum_{j=1}^{n}(y(t+j)-u(t+j))^2. \tag{6}$$

Analyzing the information in Table 1, the best test according to the training MSE is test 8. The best test according to the MSE for the prediction of one step forward is test 7. The best test for the prediction of five-step forward is test 3. Finally, the test with an optimum balance is test 7, with ten delays and five neurons in the hidden layer.

7. Analysis of results

Considering a training with ten delays and five neurons in the hidden layer corresponding to test 7, as shown in Figure 7, the actual data curve, and the prediction of a forward data.

In Figure 7, predicted data are closer to the actual data because prediction is dependent on the immediately backward data. However, this kind of prediction is not useful enough for all the market participants because it does not allow them to anticipate sudden market movements and generates imminent equal risk in a sell or buy operation. Also, another executed test is to calculate the prediction of several forward data of the series using the closed loop of the NARX neural network, feedbacking it with the output predictions. Figure 8 shows the projection of ten forward data. It can be observed that the predictions with ten forward data present a slope that helps to identify the market tendency in the medium term. This information is essential for market operators because it allows anticipating the market movement a couple of hours before.

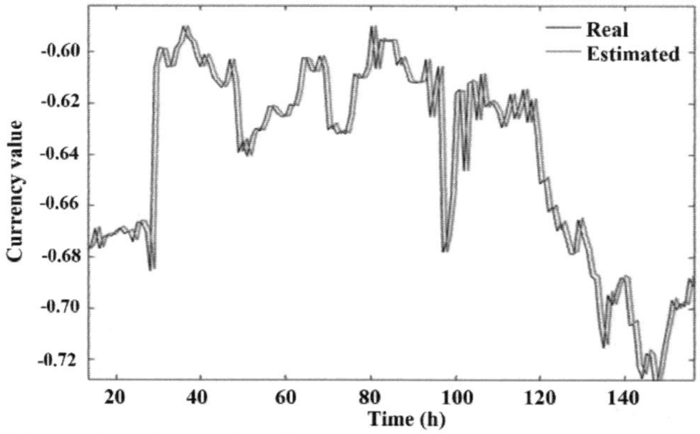

Figure 7. Price estimation curve.

With these values, an additional test is executed; calculate the predictions' slopes for five steps forward and their alignment in the same time window to verify if the information allows visualizing relevant characteristics as seen in Figure 9; the prediction slopes give information about the market momentum.

Also, this can be a strategy for researchers and investors when taking an order to the market since there is information that is not possible to see directly; for example, in Figure 9, the optimistic predictions indicate that the market still has an optimistic movement that can be taken as buy operations. On the other hand, in pessimistic predictions, the starting prices suffer a notorious fall. In other words, this would be a good indicator that the market will make a deep retracement or change in trend, allowing open a buy operation; to calculate the predictions slope, Equation (7) [13] is used.

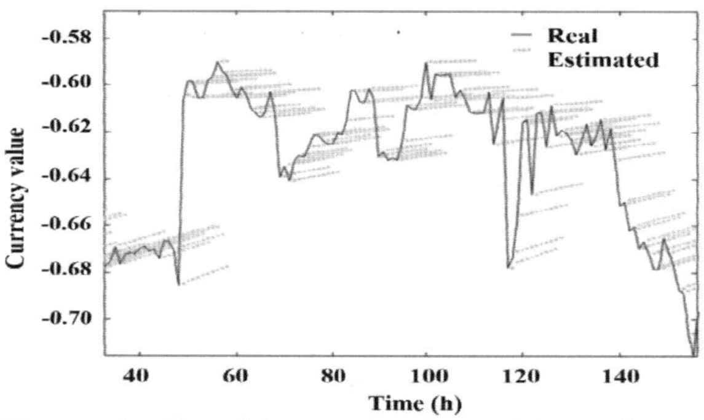

Figure 8. Closed-loop estimation with ten future data.

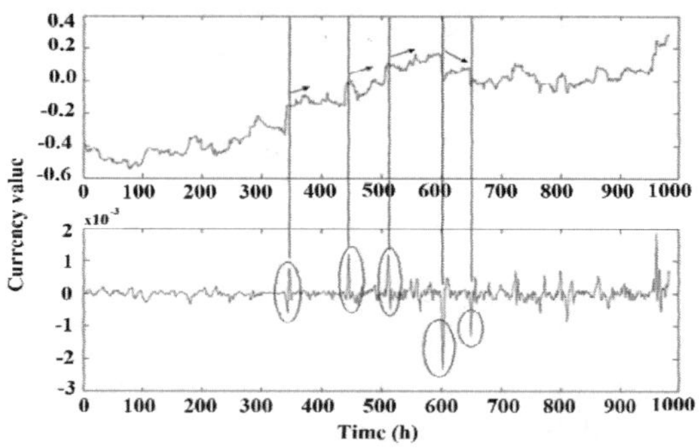

Figure 9. Prediction slopes, confirmation of market tendency.

$$mp = \frac{(y(t+n)-y(t-n))}{n-1}. \quad (7)$$

8. Conclusions

The currency market presents a nonlinear chaotic behavior, and when it is analyzed as time series of this characteristic made it possible to apply prediction techniques with good results. The recurrent neuronal networks can create time series models like those presented by the foreign exchange market.

The forward prediction does not provide valuable information to the investor due to normal price swings. Thus, using a closed loop of the NARX neural networks allows the creation of predictions of several steps forward; furthermore, analyzing the slopes of these predictions gives information about future market direction. When predictions are executed based on time series, the currency market can generate unexpected fluctuation products of impact economic news, making it difficult to use a computational technique without information to get a good prediction. Therefore, it is recommended to extract this data series that generates distortions in the models.

The use of recognition neural networks combined with prediction techniques will be a helpful strategy for a foreign exchange market participant. The previous results allow to affirm that by means of the recognition neural networks it will be possible to identify the patterns that result from the physical experimentation to clarify many of the models of theoretical physics.

References

[1] Kamruzzaman J, Sarker R A 2003 Forecasting of currency exchange rates using ANN: a case study *International Conference on Neural Networks & Signal Processing* (Nanjing: IEEE) p 793
[2] Petropoulos A, Chatzis S P, Siakoulis V, Vlachogiannakis N 2017 A stacked generalization system for automated FOREX portfolio trading *Expert Syst. Appl.* **90** 290
[3] Nerrand O, Personnaz L, Drefus G 1993 Non-linear recursive identification and control by neural networks: a general framework *Proceedings of the European Control Conference* (Groningen: IEEE) p 93
[4] Kuan C M, Liu T 1995 Forecasting exchange rates using feedforward and recurrent neural networks *J. Appl. Econom.* **10** 347
[5] Bernaschi M, Castiglione F 2001 Design and implementation of an immune system simulator *Adv. Complex Syst.* **4** 303
[6] Zheng J 2015 Forecast of opening stock price based on Elman neural network *Chem. Eng. Trans.* **46**: 565
[7] Sen J, Arup K 2012 Artificial neural network model for forecasting the stock price of Indian IT company *Proceedings of the Second International Conference on Soft Computing for Problem Solving* (Verlag: Springer India) p 1153
[8] McCluskey P 1993 *Feedforward and Recurrent Neural Networks and Genetic Programs for Stock Market and Time Series Forecasting* (Rhode Island: Brown University)
[9] Tino P, Schittenkopf C, Dorffner G 2001 Financial volatility trading using recurrent neural networks *IEEE Trans. Neural Networks* **12(4)** 865
[10] Vukovic D, Vyklyuk Y, Matsiuk N, Maiti M 2020 Neural network forecasting in prediction Sharpe ratio: evidence from EU debt market *Physica A* **542** 123331
[11] Kolen J F, Kremer S C 2001 *A Field Guide to Dynamical Recurrent Networks* (New York: Wiley-IEEE)
[12] Ni L, Li Y, Wang X, Zhang J, Yu J, Qi C 2019 Forecasting of forex time series data based on deep learning *Procedia Comput. Sci.* **147** 647
[13] Mauricio J A 2007 *Análisis de Series Temporales* (Madrid: Universidad Complutense de Madrid)

Journal of Physics: Conference Series 2118 (2021) 012018 doi:10.1088/1742-6596/2118/1/012018

Energy lost in endothermic reactions in the production of clay blocks in a continuous kiln

G Guerrero Gómez[1], N Afanador García[2], and C Nolasco Serna[3]

[1] Grupo de Investigación en Tecnología y Desarrollo en Ingeniería, Universidad Francisco de Paula Santander, Seccional Ocaña, Colombia

[2] Civil Engineering Research Group, Universidad Francisco de Paula Santander, Seccional Ocaña, Colombia

[3] Grupo de Investigación de la Facultad de Educación, Artes y Humanidades, Universidad Francisco de Paula Santander, Seccional Ocaña, Colombia

E-mail: gguerrerog@ufpso.edu.co, nafanadorg@ufps.edu.co

Abstract. In Ocaña, Norte de Santander, Colombia, the production of ceramic materials is done in a traditional and empirical manner, generating heat losses, low productivity and product quality, and increased emissions of polluting gases into the environment. A virtual instrument for temperature data acquisition was developed and the firing process was monitored in two loading doors of a Hoffman kiln. 29040 blocks were produced, and 1370.76 kg of pulverized charcoal were consumed. The firing process lasted sixty-two hours and the virtual instrument was programmed to record data every 5 minutes. The energy supplied to the brick kiln was 340.16×10^6 kJ and the heat due to loading of the products and heat lost in the endothermic reactions of the clay was 107.71×10^6 kJ and 105.71×10^6 kJ respectively, representing 31.66% and 31.08% of the energy supplied. The results have made it possible to establish trends in the temperatures and energy consumed in the endothermic reactions in the clay. This will allow the implementation of coal quality and grinding procedures, increasing energy efficiency, and reducing gas emissions into the environment, thus avoiding acute respiratory diseases.

1. Introduction

The ceramics sector in Colombia, lacks technology in its production processes [1], leading to poor combustion processes that generate environmental and health problems due to polluting emissions [2], causing unnecessary costs for the companies [3]. An inadequate combustion process requires the use of larger amounts of fuel [4] and exposes companies to fines for failing to meet environmental standards [5].

In the municipality of Ocaña, Norte de Santander, Colombia, the production of ceramic materials is done in an empirical and artisanal [6] manner. Kilns are built with non-insulating materials that generate heat losses [7], have low production and thermal efficiency [8]. Also, they do not have controls in the extrusion and firing processes [9] and their products do not meet quality standards [10].

In the manufacturing process of ceramic products in the city Ocaña, Colombia, should strategies and control systems be implemented to reduce energy consumption and gas emissions into the environment, as well as to meet product quality standards.

The objective of the research is to monitor the temperatures of the combustion product gases with a virtual temperature data instrument [11], which allows the evaluation of the energy lost in the

Journal of Physics: Conference Series **2118** (2021) 012018 doi:10.1088/1742-6596/2118/1/012018

endothermic reactions of the clay in order to implement control systems that improve the combustion of the firing process.

2. Methodology

The study used a Hoffman continuous kiln of "ladrillera Ocaña", Colombia, [12], which produces H-10 horizontally perforated units, composed of two rows with 24 chambers or doors for product loading, and 5 stacks of 620 blocks are loaded in each door, which are separated by 60 cm between them.

2.1. Coal preparation and injection system

The fuel used is coal, which is stored in a storage yard and prepared by hammer mills into tiny particles, which are conveyed to the top of the kiln by a conveyor belt system. Fuel injection begins with a first coal dosifier that starts preheating the brick located in two chambers before the second coal dosifier passes through, re-burning the brick. Coal is injected at a rate of 8 kg to 12 kg of coal per minute and coal consumption is 170 tons/month [13].

2.2. Temperature data acquisition system and temperature profiles

The acquired temperature data was achieved due to the design, programming, validation, testing, configuration and visualization of a temperature data acquisition system, based on the DAQ assistant software executable of the compact RIO embedded system. In the data acquisition process, an NI 9213 I/O acquisition card was used, which is assembled to the NI cDAQ-9184 chassis, allowing the analog-digital processing of the information received from the installed thermocouples, to be stored through the Lab view software in the acquisition report to finally generate the temperature profiles in the kiln firing process. Type K bulb thermocouples with ceramic insulation were used to measure the temperature of the products and combustion gases.

2.3. Energy supplied by the fuel

This is the energy supplied due to the combustion of the coal used in the firing process, in the evaluation of this heat, the following expression is used, see Equation (1) [14].

$$Q_H = P_c * m_c, \tag{1}$$

where Q_H= heat input in (kJ), P_c= calorific value of coal in $\left(\frac{kJ}{kg}\right)$, and m_c = fuel mass in (kg)

2.4. Heat per load of the material to be fired

This is the heat required by the products to carry out the firing process, and the following expression is used to evaluate this heat, see Equation (2), Equation (3) and Equation (4) [14].

$$Q_{cm} = Q_{cm\ L1} + Q_{cm\ L2}, \tag{2}$$

$$Q_{cm\ L1} = m_{se\ L1} * C_{om} * (T_{máx} - T_{mín}), \tag{3}$$

$$Q_{cm\ L2} = m_{se\ L2} * C_{om} * (T_{máx} - T_{mín}), \tag{4}$$

where Q_{cm}= heat per load to be fired in (kJ), $Q_{cm\ L1}$= heat per load to be fired of H10x30x20 brick in (kJ), $Q_{cm\ L2}$ = heat per load to be fired of H10x40x20 brick in (kJ), $m_{se\ L1}$ = dry mass of H10x30x20 bricks in (kg), $m_{se\ L2}$= dry mass of H10x40x20 bricks in (kg), C_{om}= specific heat of the brick material in $\left(\frac{kJ}{kg-K}\right)$, $T_{máx}$= material outlet temperature in (°C), and $T_{mín}$= material inlet temperature in (°C).

2.5. Heat to remove moisture from the material

This is the heat required to evaporate the water from the process, which has remained after drying, in the evaluation of this heat, the following expression is used, see Equation (5), Equation (6), and Equation (7) [14].

$$Q_{wm} = Q_{wm\,L1} + Q_{wm\,L2} \,, \quad (5)$$

$$Q_{wm\,L1} = m_{t\,L1} * y_{m\,L1} * (h_g - u_m), \quad (6)$$

$$Q_{wm\,L2} = m_{t\,L2} * y_{m\,L2} * (h_g - u_m), \quad (7)$$

where Q_{wm} = heat to remove the moisture from the material to be fired in (kJ), $Q_{wm\,L1}$ = heat to remove moisture from H10x30x20 bricks in (kJ), $Q_{wm\,L2}$= heat to remove moisture from H10x40x20 bricks in (kJ), $m_{t\,L1}$= total mass of H10x30x20 bricks in (kg), $m_{t\,L2}$ = total mass of H10x40x20 bricks in (kg), $y_{m\,L1}$= moisture of H10x30x20 brick in (%), $y_{m\,L2}$= moisture content of H10x40x20 brick in (%), h_g= enthalpy of water vapor in $\left(\frac{kJ}{kg}\right)$, and u_m= internal energy of the water inside the material at a temperature of 20 °C in $\left(\frac{kJ}{kg}\right)$. To evaluate the heat required to evaporate the water from the products, the enthalpy of the water vapor at a temperature of 150 °C and the internal energy of the material to be fired at 20 °C were taken [15].

2.6. Heat required for chemical decomposition of clay

This is the heat required for the chemical decomposition of the carbonates present in the clay, the following expression is used in the evaluation of this heat see Equation (8), and Equation (9), and Equation (10) [14].

$$Q_{dq} = Q_{dq\,L1} + Q_{dq\,L2}, \quad (8)$$

$$Q_{dq\,L1} = m_{se\,L1} * C_{dq}, \quad (9)$$

$$Q_{dq\,L2} = m_{se\,L2} * C_{dq}, \quad (10)$$

where Q_{dq}= heat required for the decomposition of the clay in (kJ), $Q_{dq\,L1}$= heat required for the clay decomposition of H10x30x20 bricks in (kJ), $Q_{dq\,L2}$= heat required for the clay decomposition of H10x40x20 bricks in(kJ), $m_{se\,L1}$ = dry mass of H10x30x20 bricks in (kg), $m_{se\,L2}$= dry mass of H10x40x20 bricks in(kg), and C_{dq} = specific heat from combustion of carbonates $\left(\frac{kJ}{kg}\right)$. To evaluate the heat required to break down the molecular structure of the clay and release the water from the combination of materials was evaluated at a temperature of 600 °C [15].

2.7. Heat to remove moisture from coal

This is the heat necessary to remove the moisture from the carbon, the following expression is used in the evaluation of this heat, see Equation (11) [14].

$$Q_{wc} = m_c * y_c * (h_{ga} - u_c), \quad (11)$$

where Q_{wc} = heat to remove moisture from coal in (kJ), m_C= mass of coal consumed in each period of time in (kg), y_C= coal moisture in (%), h_{ga} = enthalpy of water vapor at the temperature of the stack gases in $\left(\frac{kJ}{kg}\right)$, and u_C= internal energy of water in coal in $\left(\frac{kJ}{kg}\right)$.

2.8. Heat to remove water formed during combustion
This is the heat absorbed by the water formed by the combustion of hydrogen, which is heated, evaporated, and superheated to the temperature at which it exits with the fumes, the following expression is used in the evaluation of this heat, see Equation (12) [14]:

$$Q_{ac} = m_{cu} * m_{ac} * (h_g - u_m)\,, \tag{12}$$

where Q_{ac}= heat to remove water formed in the combustion in (kJ), m_{cu}= mass of coal consumed in each period of time in (kg), m_{ac}= mass of water formed in the combustion in $\left(\frac{kg\ H_2O}{kg\ Comb}\right)$, h_g= enthalpy of water vapor throughout the process in $\left(\frac{kJ}{kg}\right)$. and u_m= internal energy of the water inside the material at a temperature of 20 °C in $\left(\frac{kJ}{kg}\right)$; for the elemental composition of coal, emission factors are chosen for Colombian fuels, as listed by the "Unidad de Planeación Minero Energética", Colombia [16].

2.9. Heat to remove moisture in the air
This is the heat required to remove the natural moisture present in the air, which depends on the atmospheric conditions of the location, the following expression is used to evaluate this heat, see Equation (13) [14].

$$Q_{wa} = m_{wa} * m_{rm} * m_{cu} * (h_g - h_{wa}), \tag{13}$$

where Q_{wa}= heat to remove moisture in the air(kJ), m_{wa}= air humidity mass in $\left(\frac{kg\ H_2O}{kg\ Aire}\right)$, m_{rm}= dry air mass in $\left(\frac{kg\ Aire}{kg\ Comb.}\right)$, m_{cu}= mass of coal consumed in each period of time in (kg), h_g= enthalpy of water vapor in $\left(\frac{kJ}{Kg\ de\ agua}\right)$, and h_{wa}= enthalpy of water in $\left(\frac{kJ}{kg\ de\ agua}\right)$.

3. Results and discussions
27060 H 10x30x20 bricks with an average moisture content of 6.16% and 1980 H 10x40x20 bricks with an average moisture content of 2.95% were fired using 1370.76 kg of pulverized charcoal.

3.1. Virtual instrument validation
The data acquisition software was programmed to record temperatures every 5 minutes and the monitoring lasted sixty-two hours, the thermocouple recorded 744 data. The temperature profile generated from the combustion gases and theoretical clay firing curve can be seen in Figure 1.

The comparison of the theoretical temperature curve established by Munier, in Chaleur et Industries was carried out using a thermodilatometric analysis of the clay [17] with the actual firing curve of the clay obtained from the temperature acquisition. As the firing of the products is carried out due to heat transfer by radiation, the curve of the combustion gases is compared with the theoretical firing curve of the clay [17], in which there is a mismatch during the first hours of preheating in the products. On the other hand, the gases did not reach 400 °C during the first 6 hours of firing. As a result, the clay's water extraction period was extended, causing an increase in the propagation of water vapor, expansion of the clay pastes and release of a greater amount of smoke.

In the following 6 hours, the decomposition phase of the organic matter takes place and the clay curve used shows a high heating rate, reaching its maximum slope at 600 °C, causing the pieces to crack on the faces and edges most directly exposed to the fire. The maximum temperature of the gases is 800 °C, making the maturation phase of the clay complete. Finally, the temperature of the gases begins to decrease, allowing the products to begin to cool, gradually lowering their temperature and withstanding the stresses produced by a rapid cooling process.

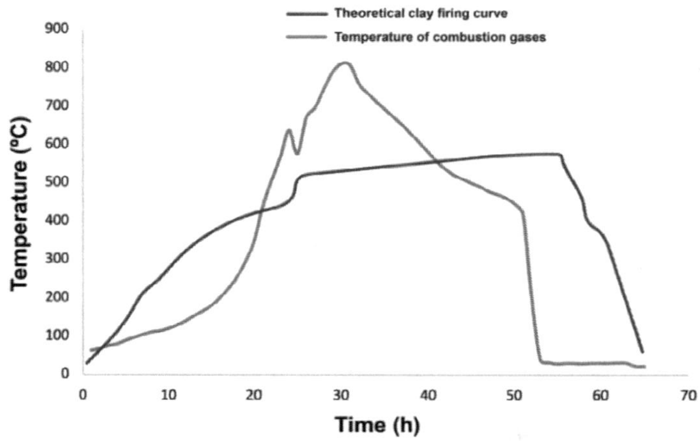

Figure 1. Temperature of combustion gases and theoretical clay firing curve.

3.2. Energy evaluation in the kiln

For the evaluation of the heat to remove water formed during combustion, the elemental composition of coal was used, taking into account the emission factors of Colombian fuels, related by the "Unidad de Planeación Minero Energética", Colombia [16]. The most important contraction of the clay in the firing process occurs between 800 °C and 900 °C [17], but although the gas reaches the maximum temperature of 800 °C, it does not remain uniform throughout the firing process. As a result, differences in contraction occur between the coldest and hottest areas of the products, causing cracks.

In the Figure 2, of the 105.71x10^6 kJ consumed by the endothermic reactions of the clay, 23.31x10^6 kJ were consumed to remove the moisture from the material to be fired which means 6.85% of the energy supplied. 58.67x10^6 kJ were used in the decomposition of the clay to remove moisture from the material to be fired corresponding to 17.24%. 1.03x10^6 kJ were used to remove moisture from the charcoal with 0.30%, 12.96x10^6 kJ to remove water formed in the combustion with 3.81%, and 9.43x10^6 kJ to remove moisture from the air with 2.77% of the energy supplied.

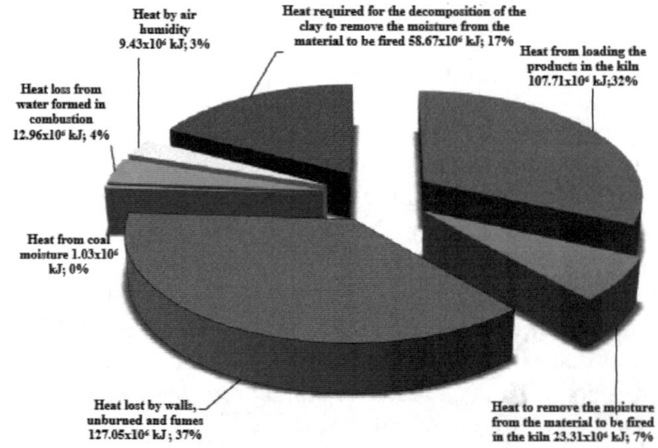

Figure 2. Energy balance in the kiln.

Journal of Physics: Conference Series **2118** (2021) 012018 doi:10.1088/1742-6596/2118/1/012018

In comparison with an energy study in a beehive kiln [18], the energy consumed by the batch in the Hoffman kiln with respect to the beehive kiln was 107.71x10^6 kJ and 7.54x10^7 kJ and the energy consumed by the endothermic reactions of the clay was 105.71x10^6 kJ and 7.30x10^7 kJ corresponding to 31.08% and 12.71% of the energy supplied respectively.

4. Conclusions

The mismatch between the gas temperature curve and the theoretical firing curve of the clay can lead to a real acid rain on the dry material, hollowing out its entire structure and staining the surfaces exposed to the gases. The increase in the temperature of the combustion gases results in a high heating rate in the products, with a maximum slope at 600 °C, which occurs as the coal dosifier approaches the material. The recorded temperature increases above 20 °C per hour for the firing stage, affects the quality standards of the products.

The temperature of the gases must be gradually increased to carry out the firing process of the products. Likewise, the cooling rates must be moderate. Otherwise, there will be differences in contractions between the coldest and hottest areas of the stack, which would generate dangerous tensile stresses and cracks. The energy percentage of the endothermic reactions of the Hoffman kiln with respect to the beehive kiln was 31.08% and 12.71% respectively of the total energy supplied, which corresponds to a higher energy consumption of 32.71x10^6 kJ in the Hoffman kiln with respect to the beehive kiln.

Therefore, one measure to mitigate the indirect loss attributable to coal supply is to implement a quality procedure and minimum requirements for coal purchase and to implement coal grinding that leads to an adequate granulometry.

References

[1] Departamento Nacional de Planeación 2007 *Agenda Interna para la Productividad y la Competitividad: Norte de Santander* (Colombia: Departamento Nacional de Planeación)

[2] Sánchez Molina J, Gelves Díaz J F, Ramírez R P 2012 Implementación de un sistema tipo scada para mejorar los procesos de secado y cocción de la ladrillera Sigma Ltda *Revista Colombiana Tecnologías Avanzada* **2(20)** 80

[3] Santiago D, Cañizares H 2012 *Diagnóstico de los Procesos de Combustión Ejecutados en la Industria Cerámica en el Área Metropolitana de San José de Cúcuta* (San José de Cúcuta: Universidad Francisco de Paula Santander)

[4] Guerrero G, Espinel E, Velásquez T 2017 Análisis isocinético y corrección a condiciones de referencia en horno a cielo abierto en el municipio de Ocaña, Norte de Santander *Revista Ingenio* **14** 43

[5] Ministerio de Ambiente, Vivienda y Desarrollo Territorial 2008 *Normas y Estándares de Emisión Admisibles de Contaminantes a la Atmósfera por Fuentes Fijas, Resolución 909* (Colombia: Ministerio de Ambiente, Vivienda y Desarrollo Territorial)

[6] British Standards Institutions (BSI) 1987 *Normas Para el Cálculo del Balance Térmico en Hornos de Ladrillo, British Standards, BS 476-13:1987* (United Kingdom: British Standards Institutions)

[7] Jahn T, Dadam A, Nicolau V 2002 Influencia da temperatura e velocidade de queima nas propriedades de tijolos comun *Anais do 46° Congresso Brasileiro de Cerâmica* (São Paulo: Associação Brasileira de Cerâmica)

[8] Peña R G, Peña Q J, Gómez T M 2015 Determinación experimental de la conductividad térmica efectiva en bloques extinguidos de arcilla roja *Ciencias en Desarrollo* **5** 15-20

[9] García Rincón J G, Arévalo Toscano J, Guerrero Gómez G 2015 Análisis exergético de un horno de lecho fijo en la producción de panela *Revista Colombiana Tecnologías Avanzada* **1(25)** 32-3

[10] Instituto Colombiano de Normas Técnicas y Certificación (ICONTEC) 2000 *Unidades de Mampostería de Arcilla Cocida. Ladrillos y Bloques Cerámicos, Norma Técnica Colombiana, NTC 4205* (Colombia: Instituto Colombiano de Normas Técnicas y Certificación)

[11] Guerrero G G, Marrugo D, Gómez J 2015 Desarrollo de instrumento virtual enfocado en la adquisición de datos para generar perfiles de temperatura en hornos *Revista Ingenio* **8** 47

[12] Herrera P, *et al.* 2011 *Caracterización de los Hornos Usados en la Industria Ladrillera* (Bogotá: Programa Eficiencia Energética en Ladrilleras Artesanales)

[13] Guerrero G, Espinel E, Flórez E 2017 Medición de emisiones atmosféricas en hornos a cielo abierto en el Municipio de Ocaña, Norte de Santander *Revista Ingenio* **14** 53

[14] Jácome S 2015 Evaluación *Termodinámica del Proceso de Cocción y Análisis de Gases en Hornos a Cielo Abierto y Hoffman en Ocaña* (Ocaña: Universidad Francisco de Paula Santander)
[15] Cengel Y, Boles M 2012 *Termodinámica* (México: McGraw-Hill)
[16] Academia Colombiana de Ciencias Exactas, Físicas y Naturales (ACCEFYN) 2003 *Factores de Emisión de los Combustibles Colombianos* (Colombia: Academia Colombiana de Ciencias Exactas, Físicas y Naturales)
[17] Guerrero G 2017 *Hornos de Combustión: Alternativas de Reducción de Emisiones e Incremento de Rentabilidad* (Bucaramanga: Universidad de Santander)
[18] Meneses J, Vera L 2003 *Evaluación de las Pérdidas de Energía en los Hornos Tipo Colmena de Tiro Invertido en la Ladrillera Cúcuta* (San José de Cúcuta: Universidad Francisco de Paula Santander)

Attitudes towards physics. A study with high school students from the Colombian context

C A Hernánez-Suarez[1], A A Gamboa-Suárez[1], and O J Suarez[2]
[1] Facultad de Educación, Artes y Humanidades, Universidad Francisco de Paula Santander, San José de Cúcuta, Colombia
[2] Facultad de Ingeniería, Universidad Autónoma de Colombia, Bogotá, Colombia

E-mail: cesaraugusto@ufps.edu.co

Abstract. Science education should promote the development of competent students in science and develop their interest in science; however, although the role and relevance of science within society is evident; the interest towards it is not conscious, being the negative attitudes of the students the main problem in the teaching of physics, which translates into a deficient knowledge and lack of scientific vocation. This report aims to diagnose the attitudes of students in high school physics courses. The study was developed through a non-experimental, descriptive research design, obtained from the application of a Likert scale that evaluates attitudes towards physics, adapted and validated to the context of students in an educational institution in Colombia. The results indicate that the attitudes of the students are slightly favorable, which is an empirical precedent to continue with the studies on attitudes towards physic. This should be considered to enhance the learning processes in these students, especially for those who have their first approach to physical concepts and give importance to attitudes when teaching.

1. Introduction

Science is one of the decisive areas for the development of society, which requires people to develop knowledge, skills and attitudes that enable them to solve real problems. However, although its role and relevance are evident, interest in science is not yet widespread in sectors outside the scientific and research spheres. This phenomenon extends to the educational context, especially at the elementary and middle school levels. Several studies have detected that the lack of interest and negative attitudes of students towards science is a problem in science education today, which translates into deficient knowledge of science [1] and a lack of scientific vocation among students in science, technology, engineering, and mathematics (STEM) programs, which are necessary for these disciplines to maintain their activity in progress.

Students' interest in studying science programs has declined in recent years [2]. The factors are diverse, but some studies suggest that students' attitudes towards science may be an important reason [3]. The Programme for International Student Assessment (PISA) has confirmed the deficiency of science education, especially in basic education, and one of its causes is that its affective component has been ignored [4-7]; therefore, it is necessary to promote positive attitudes towards science and science learning [8].

It has been shown that attitudes towards science impact and are impacted by educational experiences, with traditional teaching having the most negative impact on attitudes towards physics [9,10]. The educational value and importance of science, especially physics, in educational processes is the essence and central reason for its incorporation into academic programs. The formation and development of students' scientific attitudes is fundamental; because, faced with certain apparently complex contents, students begin to show an unfavorable attitude, as they feel incapable of overcoming the challenges inherent to the academic demands of physics. The development of the cognitive, affective, and behavioral components as pedagogical dimensions of attitudes towards physics in students leads to a better attitude to overcome their own deficiencies in learning and to obtain better achievements during their education.

This study is important because it will provide information about students' attitudes, which will allow teachers to guide the physics subject so that they have a greater impact on educational practice. It is relevant because secondary education is strategic within the education system as it is responsible for guaranteeing the appropriate transition to higher education [11]. Given the need to promote better attitudes towards physics in high school students, the question arises: What are the attitudes that students show towards science at high school levels? From this question it follows that the objective of the study was to diagnose the attitudes of high school students towards physics.

2. Attitudes towards science

A person's attitude is important in determining their interest, attention, and reaction to related subjects, for example, in the case of students towards science. In the case of Colombia, the science standards aim for students to develop the scientific skills and attitudes required to explore phenomena and to solve problems [12]. One of the difficulties in the study of attitudes is the lack of clarity in the definition of the object of attitude, which leads to inadequate interpretations of the results of research on the subject [13]. It is necessary to differentiate attitudes towards science, its teaching, learning, the teacher, the classroom environment, the didactic strategy, the contents, among others, so it is convenient to make some clarifications in this regard.

In this study, the definition of attitudes towards science was chosen as the dispositions, tendencies, or inclinations to respond to all the elements (actions, people, situations, or ideas) involved in learning science. This definition includes elements such as liking science classes, preference for science careers, science as an institution and specific science topics [4]. The dispositions or tendencies referred to in the definition can be operationally assessed through the cognitive, affective, and behavioral components towards a previously delimited attitudinal object. Attitude measurement instruments are designed, with Likert-scale items, but difficulties have been encountered at the conceptual (validity) and methodological (reliability) levels, due to the definition of the different science-related attitudinal objects that can be measured, so it is necessary to choose instruments backed by rigorous research [14].

3. Materials and method

At a general level, it is a descriptive quantitative study; the research subjects were students who study physics in grades 10 and 11 of secondary education who were characterized by attitudes towards physics class by means of a questionnaire that used a Likert scale.

3.1. Level and type of study

The study is a descriptive quantitative study, which aims to diagnose the attitudes towards physics of a group of secondary school students. Frequency is characterized by the grouping of responses.

3.2. Sample

The sample consisted of 83 students taking the physics course in secondary education (grades 10 and 11) in an educational institution located in the department of Norte de Santander, Colombia, where 43 males participated voluntarily with an average age of 18.9 ± 4.71 years and 40 females with an average age of 20.2 ± 4.21 (mean $\pm$ SD). The selection of the groups was done using purposive sampling, because the target groups are formed through entry mechanisms outside the control of the teaching staff; by grade the distribution corresponds to 50 (60.2%) students for grade 10 and 33 (39.8%) students for grade 11.

3.3. Instrument
The questionnaire used is an adaptation of the Likert-type tests assessing attitudes towards physics class [15] and towards chemistry and physics class [16] with the same 8 dimensions, each with 10 statements, for a total of 80 items with answers ranging from strongly disagree (1) to strongly agree (5); the dimensions are described in Table 1, which is taken from Muñoz and collaborators [17]. For the adaptation to the context, a pilot test was carried out in which Cronbach's alpha for the dimensions fluctuates between 0.79 and 0.85, which gives it an internal consistency, since, for instruments of this type, an approximate value above 0.70 is accepted.

Table 1. Dimensions and definition of the Likert scale for assessing students' attitudes towards physics.

Dimension	Definition
Group work	Activities that students carry out as a group to obtain a product
Individual work and homework	Activity to be carried out by the student in a personal, autonomous, and individual way for the development of their learning in physics class
Practical laboratory work	Practical work as any laboratory/experimental activity carried out in all science classes
Interests for your future	Personal linkage of the subject regarding the valuation he/she presents towards science to be considered in future life projects
Teacher's influence about physics	The teacher is understood to be a stimulating factor towards attitudes; therefore, it deals with the teacher-student (individual) linkage and their relationship in terms of classroom didactics and content
Difficulty in learning physics	The content and perceived difficulty of the student's effort to achieve in the subject is covered
Relationship between everyday life and the physics course	The aim is to find the relationship that the student has with the subject based on the link between the subject and the natural/observable world and the contents of the subject
Social importance, importance of science and scientists	It is understood that attitudes are fostered by a positive perception of scientists and science

3.4. Procedure
Authorization was requested from the directors of the educational institution to obtain the information, and once obtained, the instrument was placed in digital format on the educational institution's platform during the timetable assigned to each group; both teachers and students were informed of the purpose of the study and expressed their willingness to participate. With the help of the teachers in charge of each of the groups, the students responded voluntarily during the assigned dates, which correspond to the first semester of 2021. Only those students who completed the questionnaire were considered.

3.5. Data processing
The arithmetic mean per item and the overall score per test dimension were calculated using the Microsoft Excel spreadsheet. For the interpretation of the numerical scale, it follows the methodological guidelines of the Likert scales [18], which means that scores below 3.0 indicate an unfavorable attitude and, for scores above 3.0, a favorable attitude to physics will be considered as a favorable attitude to physics.

4. Results
Next, the results obtained from the application of the test of attitudes towards physics class and its respective dimensions to high school students are presented.

4.1. Attitudes towards physics among secondary school students
The overall results of the attitudes expressed by secondary school students and the descriptive comparisons by grade for the dimensions of the instrument are shown in Figure 1; from Figure 1,

students show a favorable attitude towards physics (3.3), especially quite favorable attitudes towards individual work and homework (3.6), work in laboratory practice (3.6) and the relationship between everyday life and the subject of physics (3.5), and less favorable attitudes towards the difficulty of learning physics and interests for the future (3.0).

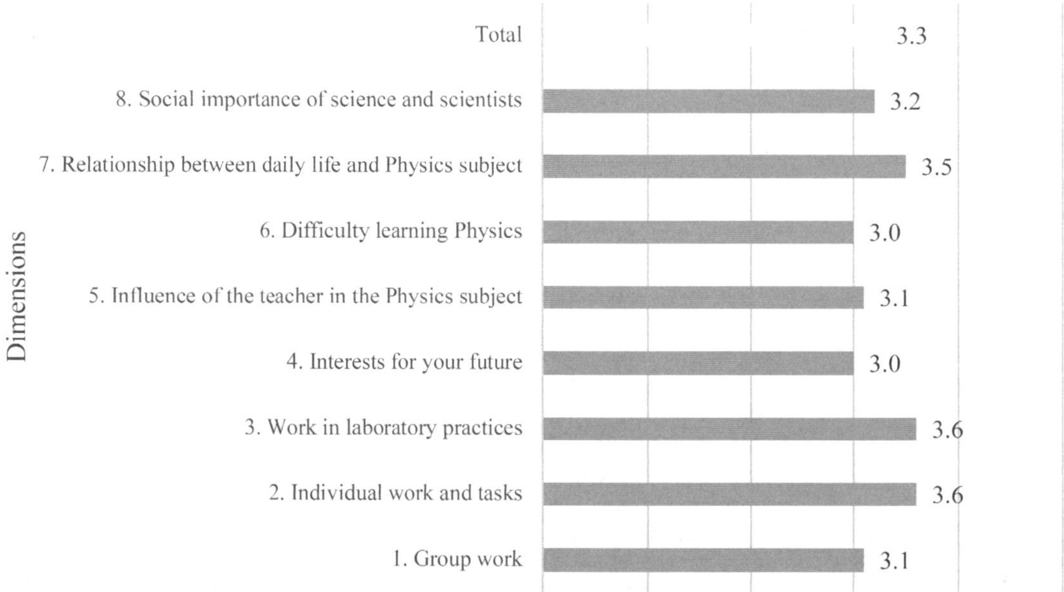

Figure 1. Overall average of scores per dimension of the instrument.

4.2. Attitudes towards physics in students by grade level

In the following, we will explore students' attitudes towards physics by grade and identify possible differences between them. This comparison is made because the aim is to show how attitudes evolve from one grade to the next at the secondary school level; considering the information in Table 2, there is no difference between the average score of grades 10 and 11.

Table 2. Overall comparison of means between 10th and 11th grades.

Grade	Arithmetic mean
10	3.2
11	3.2

Considering the values presented in Table 3, it can be indicated that there are no differences for the dimensions: interests for their future, relationship between everyday life and the physics course, and social importance of science and scientists, which indicates that there are no differences for these dimensions between grades 10 and 11; however, in the dimensions: group work, individual work and tasks, work in laboratory practices, teacher influence on the physics course, and difficulty in learning physics, differences are observed. From the previous result, the difference between students in grades 10 and 11 for the dimension difficulty in learning physics stands out, since students in grade 11 show a favorable predisposition, while the attitude of students in grade 10 is below the favorable one.

High school students show a slightly favorable predisposition towards Physics class, with a value higher than 3, which is corroborated in various studies [19-22], but which contrasts with a study that agrees that the predispositions that students manifest towards physics are negative, especially in those with low academic performance [23]. The favorable predisposition of the students in this study indicates that there is a basis on which to enhance students' general interest in physics, which could be achieved with innovative strategies such as active learning [24].

This study constitutes an empirical background for further research on attitudes towards physics in students, other levels, and grades, as well as their relationship with other variables such as motivation [20], academic performance [23], gains in learning physical concepts [25].

Table 3. Comparison between 10th and 11th grade by dimension.

Dimension	Grade			
	10		11	
	Average	Sd	Average	Sd
Group work	3.2	0.77	3.0	0.66
Individual work and homework	3.6	0.71	3.5	0.69
Practical laboratory work	3.6	0.65	3.7	0.67
Interests for your future	3.0	0.82	30	0.84
Teacher's influence on the physics course	3.1	0.51	3.2	0.53
Difficulty in learning physics	2.8	0.87	3.1	0.78
Relationship between everyday life and the physics course	3.5	0.70	3.5	0.69
Social importance, importance of science and scientists	3.2	0.53	3.2	0.49

5. Conclusions

Studies of attitudes towards learning physics are relevant, as it is considered a difficult subject for students to learn. The results of this study suggest that secondary school students have a slightly favorable attitude towards physics. As for the comparison between the attitudes of Grade 10 and 11 students, no differences were found overall. There are only differences in the dimension of difficulty in learning physics, where an unfavorable attitude was observed in Grade 10 students, possibly due to the degree of satisfaction experienced by these students as it is their first approach to the subject of physics. This should be considered to enhance the learning processes in these students, especially for those who have their first approach to physical concepts and to give importance to attitudes when teaching. Finally, future studies on attitudes towards physics should include other variables such as academic performance, motivation, innovative strategies, the teacher, and preconceptions of physics, among others. In addition, it could be extended to university students of basic sciences and engineering, or in other types of populations, these being fields of study for future research.

References

[1] Matus M 2013 Actitud hacia la ciencia en estudiantes de una universidad estatal de Valparaíso *Revista de Psicología* **2(4)** 57

[2] Organización para la Cooperación y el Desarrollo Económico 2008 *Informe PISA 2006. Competencias Científicas para el Mundo del Mañana* (Madrid: Santillana)

[3] Organización para la Cooperación y el Desarrollo Económico 2006 *Evaluación de la Competencia Científica, Lectora y Matemática: Un Marco Teórico para PISA 2006* (Madrid: INECSE/MEC)

[4] García-Ruíz M, Sánchez B 2006 Las actitudes relacionadas con las ciencias naturales y sus repercusiones en la práctica docente de profesores de primaria *Perfiles Educativos* **28(114)** 61

[5] Garritz A 2009 La afectividad en la enseñanza de la ciencia *Educación Química* **20(Supl. 1)** 212

[6] Ojeda E, Delgado C, Ascanio A, Pérez M E 2011 Emociones: su impacto en la modificación de concepciones de docentes en formación en educación integral del Instituto Pedagógico de Caracas *Revista de Investigación* **35(74)** 139

[7] Cervantes R D, Reséndiz E, Correa S 2019 Análisis de los componentes actitudinales hacia la ciencia en maestros participantes en el programa de enseñanza de la ciencia *Innovación en Enfoques de Contextos Escolarizados* ed Reséndiz E, Cappello H M, García H M (Ciudad de México: Colofón)

[8] Ruiz F J 2007 Modelos didácticos para la enseñanza de las ciencias naturales *Revista Latinoamericana de Estudios Educativos* **3(2)** 41

[9] Solbes J, Montserrat R, Furió C 2007 Desinterés del alumnado hacia el aprendizaje de la ciencia: implicaciones en su enseñanza *Didáctica de las Ciencias Experimentales y Sociales* **(21)** 91

[10] Torres M 2010 La enseñanza tradicional de las ciencias versus las nuevas tendencias educativas *Revista Electrónica Educare* **14(1)** 131

[11] Jurado F 2019 Retos para la transición entre la educación media, la educación posmedia y la educación superior *Ruta Maestra* **27** 136

[12] Ministerio de Educación Nacional 2004 *Estándares Básicos de Competencias en Ciencias Naturales y Ciencias Sociales* (Bogotá: Ministerio de Educación Nacional)

[13] Parra H, Santos R 2014 *Actitud de los Estudiantes Hacia el Aprendizaje de la Física en el Tercer Año de Educación Media General. Caso: Unidad Educativa Casa Don Bosco del Municipio Naguanagua. Año Escolar 2013-2014* (Naguanagua: Universidad de Carabobo)

[14] Tecpan S, Zavala G 2015 Actitudes de profesores de Física de secundaria sobre ciencia, tecnología y sociedad *Lat. Am. J. Phys. Educ.* **9(1)** S1202:1

[15] Gómez A G 2011 *Las Actitudes Hacia la Clase de Física del Estudiantado de Secundaria; un Estudio Exploratorio Descriptivo en Instituciones Educativas de Santiago y Concepción* (Concepción: Universidad de Concepción)

[16] Boixaderas N, De la Vila J,Sanmartí N 1990 *Test de Actitudes Relacionadas con la Asignatura de Física y Química* (Barcelona: Universidad Autonoma de Barcelona)

[17] Munoz D, Quintanilla M R, Manzanilla M A 2019 Construcción y validación preliminar de un instrumento de evaluación de actitudes hacia la clase de química para estudiantes de educación secundaria *Educación Química* **30(1)** 121

[18] Hernández R, Fernández, C Baptista M P 2010 *Metodología de la Investigación* (México; Mc Graw Hill)

[19] Pinto F 2015 Un estudio de las actitudes hacia la física de los estudiantes cursantes del quinto año de educación media general mención ciencias de los liceos públicos y privados municipio Naguanagua *ARJÉ Revista de Postgrado FACE-UC* **9(16)** 103

[20] Arandia E, Zuza K, Guisasola J 2016 Actitudes y motivaciones de los estudiantes de ciencias en Bachillerato y Universidad hacia el aprendizaje de la Física *Revista Eureka Sobre Enseñanza y Divulgación de las Ciencias* **13(3)** 558

[21] Guevara J C 2018 Actitudes de los estudiantes universitarios hacia el aprendizaje de la física *Anales Científicos* **79(1)** 1

[22] Torres N, Bolívar A, Solbes J, Parada M 2018 Percepciones de estudiantes universitarios sobre su formación en física en educación secundaria *Revista U.D.C.A Actualidad & Divulgación Científica* **21(2)** 599

[23] Zeballos R M A 2019 *Estudio de las Actitudes Hacia la Física y su Relación con el Rendimiento Académico de los Estudiantes de las Escuelas Profesionales de Ingenierías, Física y Facultad de Educación de la UNSA, 2017* (Arequipa: Universidad Nacional de San Agustín)

[24] Prada R, Hernández C A, Gamboa A A 2021 Active learning, and knowledge in physics: a reading from classroom work *Journal of Physics: Conference Series* **1981(1)** 012007:1

[25] Santana J L 2018 Ganancia en el aprendizaje del concepto de fuerza y cambio en las actitudes hacia la física en estudiantes de la escuela preparatoria de Tonalá *Ciencia UAT* **13(1)** 65

Teaching of physics applied to engineering higher education by using English language: a literature review

Ó H Vargas Villamizar[1]
[1] Facultad de Ciencias Económicas y Administrativas, Universidad El Bosque, Bogotá, Colombia

E-mail: ovargasvi@unbosque.edu.co

Abstract. The use of English as a second language is an increasing trend in higher education focused on non-native speakers, particularly for engineering programs interested in becoming actors inserted into the international academic context. However, the dual condition of both teaching engineering curriculum contents, and simultaneously using English language in non-native speaking contexts, entails risks and challenges that should be deeply understood. Considering this, results from a literature review are presented, built on articles from academic databases regarding use of two approaches, English as a medium of instruction and Content and language integrated learning, in engineering at higher education institutions, with specific interest in physics teaching, because of its relevance in engineering body of knowledge. As a finding, this topic has not been sufficiently studied in academic literature, and two focuses to carry out further research are suggested to contribute to the understanding of how to teach physics by using English language in non-native speaking contexts.

1. Introduction

English language has become world's lingua franca, its influence has permeated social and cultural spheres, and the academic sector is not an exception: as of 2018, the number of refereed academic journals listed in Ulrich's Periodicals Directory reached 77,218 published in part or fully in English language [1]. In this scenario, higher education institutions in non-English speaking countries are under pressure to rapidly update their pedagogical framework by offering bilingual programs and courses to be appealing for students and high-level professors [2].

Learning a second language while studying subject content may improve motivation, learning strategies, and self-confidence in comparison to monolingual students [2]. Even though, the decision of using English in non-language courses is particularly challenging in engineering programs because of, among others, the level of mathematics and physics involved [3], and this pedagogical choice entails some problems as Hincks R [4] detected when analyzing a group of 14 English fluent (C1/B2) master's students of engineering in Sweden: she found that speaking rate in English was 23% slower in comparison to lessons in their first language.

Using a second language to teach a highly complex body of knowledge without compromising learning outcomes is a difficult task, as Airey J and Linder C [5] found in a sample of 22 undergraduate physics students at two universities in Sweden: they asked and answered fewer questions when the class was led in English, and they had more problems when simultaneously listening to professors and taking notes. In addition, dealing with using English and teaching content is highly demanding for professors, and resistance may raise against this new dual role, as Airey J [6] identified in qualitative research carried out with ten physics lecturers from four Swedish universities involved in English taught courses, and as a result, none of them conceived themselves as language teacher.

In this scenario, nevertheless, Thøgersen and Airey [7] states a small number of studies about the consequences of using English language, through observation or experimental designs, have been carried out, and Kukkonen P [8] considers that, although language acquisition and the learning of physics are topics intensively studied, there is a lack of articles about English as a second language to teach physics. The aim of this paper is to conduct a literature review process in order to identify if there is a research gap regarding how English language is used, in non-native speaking contexts, to teach Physics in engineering higher education.

2. Background

In this study two main approaches oriented to use English language while teaching subject content will be addressed: English as a medium of instruction (EMI) and content and language integrated learning (CLIL); they both aim to help students to improve their proficiency in a second language while learning disciplinary knowledge, but CLIL teaching implies an equal focus on content and language whereas EMI contexts are more interested in teaching subject contents, and language learning becomes a secondary priority [9].

2.1. English as a medium of Instruction in higher education

English as a medium of instruction is defined as "the use of the English language to teach academic subjects (other than English itself) in countries or jurisdictions where the first language (L1) of the majority of the population is not English" [10]. The increasing interest to offer EMI programs in higher education is context-dependent and related to efforts to gain prestige, to attract foreign students (due to falling of home students' applications), to counteract external sources' investment cutting, and to embrace English as a status language (particularly in research publications) [10]. Although the implementation of EMI in higher education has rapidly been growing worldwide, academic research aimed to understand its effectiveness has been lagged behind [11], and analytic research to understand classroom interaction in higher education with English as a Medium of Instruction is scarce [12].

Moreover, opposition to use EMI in class is increasing, and Başibek N, *et al.* [13] brought up teaching in English without appropriate curricula might foster feelings of confusion and insecurity between professors and students; regarding that, Başibek N, *et al.* [13] compiled six arguments against EMI approach: lower ability to understand concepts, reduced knowledge about subjects, it is time-consuming process, raising of feelings of alienation and separation, participation is classes decreases, and EMI might be seen as "a vehicle for creating an elite class" by limiting the involvement of students with no prior access to education in English.

2.2. Content and language integrated learning in higher education

CLIL is defined as "a dual-purpose approach" that combines teaching curriculum content and simultaneously a second language (L2) in an integrated manner [14]. According to Wewer T [15], CLIL's purpose is to strengthen the language repertoire by using the second language as a medium of teaching and learning contents across the curriculum, and its popularity has been thriving in the last decades in Latin America and Asia as a language instruction approach.

Among the advantages, Banegas DL and del Pozo Beamund M [14] compiled studies from experts reporting benefits in features such as motivation, intercultural awareness and linguistic development in young learners and high education students.

Nevertheless, in their research about the teaching of a physical law by using English language, Käänta L, *et al.* [16] claimed there is limited literature on physics education in a CLIL classroom, and they state, as a main result from the study, it is necessary to coordinate multilingual and multimodal practices to succeed in this dual challenge. In accordance with that, Käänta L and Kasper G [17] pointed at the difficulty to determine exactly what understanding is gained by students to future use.

3. Methodology

In order to get a better understanding on conditions related to English as a second language to teach physics in Engineering at tertiary education, a four steps literature review was carried out by turning to Scopus and ScienceDirect as bibliographic sources, and the review process and summary frameworks were designed by following Fontina, *et al.* [18] proposal; both Scopus and ScienceDirect were chosen as providers of reliable peer-reviewed articles, and an unrestricted period was considered due to study's highly specific scope. Search process was developed as follows. Firstly, literature search was based on four keywords used in conjunction, connected by the Boolean "AND":

- Keyword 1: "English as a medium of instruction" (and its variation "English medium instruction") and content language and Integrated learning (and its variation "CLIL"). Additionally, to get a wider scope, "English as a foreign language (EFL)" and "English as a second language (ESL)" were used too.
- Keyword 2: "engineering".
- Keyword 3: "physics".
- Keyword 4: "higher education" (and its variations "tertiary education", "university").

Secondly, and based on the abovementioned keywords, articles from journals were collected and duplicate ones were removed. After that, titles and abstracts for preliminary articles were carefully read to discard those ones unrelated to this study scope for reasons such as they are not oriented to pedagogical issues, or they are unconnected to either engineering or physics. Later, articles that passed the filter were read in-depth to identify any relation to the study's topic, dropping all of them with no direct connection. Finally, the remaining articles were used to write a descriptive analysis presented in discussions.

The methodology and criteria used to identify articles are concisely described in Table 1, and the number of articles identified during search process is showed in Table 2 (both based on frameworks proposed by Fontina, *et al.* [18]).

Table 1. Summary of the literature review process.

Step 1 and Step 2	Finding and evaluation of articles Electronic databases: Scopus (scopus.com) and ScienceDirect (sciencedirect.com) Search period: unrestricted time period Inclusion criteria: articles written in English language with title and abstract related to the aim of this paper, based on four keywords, and finally chosen after a full review of their contents Exclusion criteria: articles unrelated to the four keywords used in conjunction, and articles not available in English language Search string: "EMI / CLIL" AND "Engineering" AND "Physics" AND "Higher education"
Step 3	Assessment of findings Analysis phase: collection of articles from databases
Step 4	Reporting of findings Synthesis phase: relevant articles are analyzed and discussed

Table 2. Summary of the literature review search process.

	Search string	Articles identified	Articles post removing duplicates	Articles post abstract review	Articles post full text review
Total	"EMI / CLIL / EFL / ESL" AND "Engineering" AND "Physics" AND "Higher education"	130	92	15	4

4. Discussions

From an initial sample of 92 articles identified in the literature review process, 15 were chosen to be read in-depth, written from 1985 [1] to 2021[5,8,10]. After the post full text review, 11 of them were discarded because, although they all offer relevant information about challenges of using English as a second language, their contributions to understand the object of this study were not significant.

From these 11 papers, 3 of them are oriented to reflect upon students' learning process by systematizing their perceptions of language abilities and needs [19], by processing how English vocabulary is learned [20], or by comparing courses taught in a second language with conventional ones [21]. Other 4 articles are focused on discussing theoretical models and descriptors [22], exploring literacy support [23], or using strategies such as arts and humanities [24] or flipped classroom [25] in tertiary education courses. Finally, 4 papers aim to analyze specific variables like disciplinary cultures [26], ambiguity tolerance [27], supervisory feedback [28], and speaking rate [7] and how they affect learning processes.

From the sample of 15 articles, 4 of them contain relevant insights to understand the chosen topic. Furthermore, there is only one article whose title, abstract and content fully matches to the study scope, and the other three papers offer more limited contributions to the object of study.

The above-mentioned significant article was written by Tatzl D and Messnarz B [29] with the aim to investigate "the influence of English as the examination language on the solution of physics and science problems by non-native speakers in tertiary engineering education" in a degree program of aviation at the FH Joanneum University of Applied Sciences (Graz, Austria) whose students are involved in aeronautical engineering and aviation management. Non-native speakers of English, according to Tatzl D and Messnarz B [29], are leaned to be part of English taught programs as a way to get a more competitive and employable work profile. Likewise, Tatzl D and Messnarz B [29] remarks a difference between Austrian students, in comparison to other countries, regarding previous knowledge of the language, because an English level B2 is expected in prospects entering to tertiary education.

As a result, Tatzl D and Messnarz B [29] suggests that "the impact of English as a foreign language skill on mathematical problem-solving in science and physics is ambiguous, to say the least" and his investigation could not validate his initial assumption that non-native speakers would need additional time and aids to solve engineering physics problems in English; from Tatzl D and Messnarz B [29] perspective, tailor-made learning proposals should be designed to each group of learners because, apparently, a second language does not affect solving numerical problems, but achievement rates are affected if written descriptions are required [29].

The second article identified in this review stressed on teachers' beliefs about using translanguaging (coexistence of multiple languages as a communication system) as pedagogical practice [30]. The research includes semi-structured interviews, classroom observations, and analyses of teaching and assessment material from four teaching staff in Iraqi universities. As a result, Alhasnawi S [30] considers adopting a 'one-size-fits-all' perspective on EMI limits our understanding on how non-native speakers interact with English language, ignoring geopolitical and socio-cultural implications, considering both limited proficiency in English and in mathematics in the Iraqi context (for instance, one of the professors expressed he teaches 80% of physics content in Arabic instead of English). Likewise, Alhasnawi S [30] encourages developing serious research on using languages for academic purposes and pedagogical skills related to disciplinary practice.

In the third article, Reus L [31] stresses on how EMI may counteract Latin American students' "language exchange isolation". According to Reus L [31], in that region EMI is still in an early stage, but it is growing because of the interest of universities to attract foreign students and to climb positions in international rankings. This study was developed in a Chilean engineering school by using multivariate analysis of tests and final grades in two economics-related courses, and as a result, there are not negative effects on content acquisition, but the author remarks it is not possible to extrapolate this result to other areas as physics because students who enjoy economics do not necessarily enjoy math [31].

Finally, in the fourth article, written communication as a core competence for higher education students is the focus in Bylander J and Gustafsson M [32] investigation: experimental lab-report writing assignments in English in a quantum engineering course written by master's-level physics were studied, and the author considers his proposed method might be used in other STEM disciplines. Although this study is more oriented to "learning to write" activities for academic purposes, Bylander J and Gustafsson M [32] detects the implications of uneven student groups' English skills and peer-feedback in writing high quality reports in English.

According to the literature review results, the use of English as a second language to teach physics in engineering higher has not been sufficiently analyzed: The only article that fully matches to that scope, written by Tatzl D and Messnarz B [29], is focused on a specific process (examination) for a sample of proficient students (English level B2 or more) enrolled in the same undergraduate program in a high-income country (Austria). Based on this finding, two focuses to carry out further research are suggested.

The first focus is related to identify how professors modify their English as a second language approaches to teach physics in engineering according to the educational moment (lecturing, classroom discussions, testing), and the type of content, either math-oriented courses (such as theoretical physics courses) or observation-oriented ones (as Bylander J and Gustafsson M [32] point at analyzing experimental lab-reports). The alignment between language training and content skills in engineering education, according to Tatzl D and Messnarz B [29], rests on interaction between content and language instructors, so to investigate about this multidisciplinary relation would provide useful insights to improve curriculum planning and implementation.

The second focus emerges from the lack of significant studies about challenges to adapt approaches such as EMI or CLIL to teach physics in engineering programs in contexts with low (or uneven) proficiency from students and professors. Differences between these both approaches should be documented to develop research-informed pedagogical practices [12], therefore, it is recommendable to undertake specific research in how to teach physics from an engineering perspective in order to not extrapolate outcomes based on experiences from another body of knowledge (as Reus L [31] warns). Finally, further studies should be performed in medium or low-income countries because using English as a second language in these contexts demands more flexible frameworks, as Alhasnawi S [30] urges for Iraqi universities, taking into account affective factors that have not been traditionally involved in physics instruction [8].

5. Conclusion

From the literature review process, it can be concluded there is a knowledge gap regarding how English language is used, in non-native speaking contexts, to teach physics applied to engineering higher education. From 92 papers identified in the preliminary search process, and after an abstract review, 15 of them were considered suitable for a full text review, and as a result, just 4 articles were identified as sources of relevant insights, but only one of them fully fitted the topic of interest.

Based on the literature review and the descriptive analysis showed in previous chapters, it has been identified lack of experimental studies about two main educational challenges in physics: How professors manage teaching approaches such as English as a medium of instruction and content and language integrated learning when leading physics courses for engineering programs; and how English language should be applied to teach physics in engineering courses in developing countries, where both second language command and math skills are limited.

Developing curriculum adjustments to effectively teach physics relies on available information from well-documented best practices, and the lack of data-driven decisions might hinder full insertion of students and professors in both international research networks and postgraduate programs. Therefore, university physics and engineering departments must lead research regarding how English language is actually used in their different learning stages, as a way to systematize their experiences by documenting new knowledge to be shared within academic community.

Learning a second language is a motivating but frightening experience, and learning physics is a challenging endeavor itself, so combining both activities is quite difficult task. Particularly for Colombia, whose citizens have low English literacy (according to EF English proficiency index) and whose results in mathematics are lower than in comparable countries (according to the Organization for Economic Cooperation and Development), this study exhorts to perform local research (based on the two focuses mentioned in discussions chapter) about how professors and universities are dealing with English language in classrooms and labs, and what kind of solutions have been proposed (or might be proposed) by academic actors to adjust pedagogical processes to embrace English language in higher education physics courses.

References

[1] McDowell L, Liardét C L 2019 Japanese materials scientists' experiences with English for research publication purposes *Journal of English for Academic Purposes* **37** 141

[2] Rivero-Menéndez M J, Urquía-Grande E, López-Sánchez P, Camacho-Miñano M M 2017 Motivation and learning strategies in accounting: Are there differences in English as a medium of instruction (EMI) versus non-EMI students? *Revista de Contabilidad* **21(2)** 128

[3] Chowdury H, Alam F, Mustary I 2019 Development of an innovative technique for teaching and learning of laboratory experiments for engineering courses *Energy Procedia* **160** 806

[4] Hincks R 2010 Speaking rate and information content in English lingua franca oral presentations *English for Specific Purposes* **29(1)** 4

[5] Airey J, Linder C 2006 Language and the experience of learning university physics in Sweden *European Journal of Physics* **27(3)** 533

[6] Airey J 2012 "I don't teach language": the linguistic attitudes of physics lecturers in Sweden *AILA Review* **25** 64

[7] Thøgersen J, Airey J 2011 Lecturing undergraduate science in Danish and in English: a comparison of speaking rate and rhetorical style *English for Specific Purposes* **30(3)** 209

[8] Kukkonen P 2006 Toward a working theory for physics CLIL classrooms *Exploring Dual-Focused Education. Integrating Language and Content for Individual and Societal Needs* ed Björklund S, Mard-Miettinen K, Bergström M, Södergård M (Vaasa: Centre for Immersion and Multilingualism, University of Vaasa) p 147

[9] Galloway N, Ruegg R 2020 The provision of student support on English Medium Instruction programmes in Japan and China *Journal of English for Academic Purposes* **45** 100846

[10] Macaro E, Curle S, Pun J, An J, Dearden J 2017 A systematic review of English medium instruction in higher education *Language Teaching* **51(1)** 36

[11] Graham K M, Eslami Z R, Hillman S 2021 From English as the medium to English as a medium: perspectives of EMI students in Qatar *System* **99** 102508

[12] Duran D, Sert O 2019 Preference organization in English as a medium of instruction classrooms in a Turkish higher education setting *Linguistics and Education* **49** 72

[13] Başıbek N, Dolmacı M, Cengiz B C, Bür B, Dilek Y, Kara, B 2014 Lecturers' perceptions of English medium instruction at engineering departments of higher education: a study on partial English medium instruction at some state universities in Turkey *Procedia – Social and Behavioral Sciences* **116** 1819

[14] Banegas D L, del Pozo Beamud M 2020 Content and language integrated learning: a duoethnographic study about CLIL pre-service teacher education in Argentina and Spain *RELC Journal* doi: 10.1177/0033688220930442

[15] Wewer T 2017 good practices for primary-level content-based foreign language instruction identified in Finland *Applied Linguistics Journal* **19(2)** 277

[16] Kääntä L, Kasper G, Piirainen–Marsh A 2018 Explaining Hooke's law: definitional practices in al CLIL physics classroom *Applied Linguistics* **39(5)** 694

[17] Kääntä L, Kasper G 2018 Clarification requests as a method of pursuing understanding in CLIL physics lectures *Classroom Discourse* **9(3)** 205

[18] Foncina A, Introva V, Silvestri A 2021 Enabling technology for maintenance in a smart factory: a literature review *Procedia Computer Science* **180** 430

[19] Zughoul M R, Hussein R F 1985 English for higher education in the Arab world: a case study of needs analysis at Yarmouk University *The ESP Journal* **4(2)** 133

Journal of Physics: Conference Series **2118** (2021) 012020 doi:10.1088/1742-6596/2118/1/012020

[20] Zhao Y T Y, Jia Z Y, Tang Y, Jie Xiong J, Zhang Y C 2018 Quantitative Learning strategies based on word networks *Physica A: Statistical Mechanics and its Applications* **491** 898
[21] Más-Estellés J, Benlloch-Dualde J V, Gil-Vicente P, Gil-Salom D 2018 English-medium instruction groups in informatics engineering: results and comparison to conventional groups *IEEE Global Engineering Education Conference (EDUCON)* (Santa Cruz de Tenerife: IEEE) p 399
[22] Martín del Pozo M A 2015 Signposts for comprehensive knowledge of CLIL Contexts *Procedia – Social and Behavioral Sciences* **212** 99
[23] Sánchez-Pérez M M 2021 Predicting content proficiency through disciplinary-literacy variables in English-medium writing contexts *System* **97** 102463
[24] Kim J H, Nguyen N T T, Campbell R C, Yoo S, Taraban R, Reible D D 2021 Developing reflective engineers through an arts-incorporated graduate course: a curriculum inquiry *Thinking Skills and Creativity* **42** 100909
[25] Karjanto N, Simon L 2019 English-medium instruction calculus in Confucian-heritage: flipping the class or overriding the culture *Studies in Educational Evaluation* **63** 122
[26] Chang Y Y 2012 The use of questions by professors in lectures given in English: influences of disciplinary cultures *English for Specific Purposes* **31** 103
[27] Atamanova I, Bogomaz S 2014 Ambiguity tolerance as a psychological factor of foreign language communicative competence development *Procedia – Social and Behavioral Sciences* **154** 345
[28] Neupale Bastola M, Hu G 2021 Supervisory feedback across disciplines: does it meet students' expectations? *Assessment & Evaluation in Higher Education* **46(3)** 407
[29] Tatzl D, Messnarz B 2012 Testing foreign language impact on engineering students' scientific problem-solving performance *European Journal of Engineering Education* **38(6)** 620
[30] Alhasnawi S 2021 Teachers' beliefs and practices with respect to translanguaging university mathematics in Iraq *Linguistics and Education* **63** 100930
[31] Reus L 2020 English as a medium of instruction at a Chilean engineering school: experiences in finance and industrial organization courses *Studies in Educational Evaluation* **67** 100930
[32] Bylander J, Gustafsson M 2021 Improved content mastery and written communication through a lab-report assignment with peer review: an example from a quantum engineering course *European Journal of Physics* **42(2)** 025701

Analysis of the application of physics in the design and construction of architectural projects. The Eiffel Tower

J D González-Almeyda[1], E T Ayala-Garcia[2], and R Prada-Nuñez[3]
[1] Facultad de Educación, Artes y Humanidades, Universidad Francisco de Paula Santander, San José de Cúcuta, Colombia

E-mail: erikatatianaayala@ufps.edu.co

Abstract. This article studied the impact and application of physical concepts in the design and construction of the Eiffel Tower in Paris, an architectural reference that implemented physical concepts in its structural design. A documentary methodological framework was used to establish the importance of the Eiffel Tower in the universal exposition of Paris in 1989 and, to carry out the structural analysis of the work; a quantitative-descriptive approach was used for the recognition of the basic concepts of physics from architecture according to gender, through a survey as a research instrument developed under non-probability and convenience sampling, which was applied to students of Architecture of the Universidad Francisco de Paula Santander, Colombia, in order to determine the knowledge of basic physics by students. The results of this research showed that the Eiffel Tower represents a milestone in architecture where physical concepts such as tension, compression, traction, aerodynamics, and torque were applied. Finally, it was evidenced that the students who participated in this study recognize the importance of applying the basic concepts of physics in architecture; fact by which it is recommended to encourage the study of physics, to strengthen the technological component of Architecture.

1. Introduction

The industrial revolution in Europe in the 19th century brought as a consequence great change of social, economic, political and cultural order that generated transformations in the cities. The implementation of new materials such as iron consolidated a dynamization of the economy that generated an acceleration in its production and use, impacting multiple scenarios in society [1]. Consequently, towards the middle of the 19th century, with the objective of making visible the technical, technological and scientific advances of the nations, the universal exhibitions were consolidated [2], which were characterized by presenting specific large-scale architectural landmarks, which were intended to host the exhibitions and, to demonstrate the grandeur and national supremacy from the technical and the technological, The first universal exhibition corresponded to London in 1851 with the Crystal Palace [3].

Between May 6 and October 31, 1889, the universal exhibition that commemorated the centenary of the French revolution was held in Paris, this exhibition was characterized by presenting the Eiffel Tower as a landmark or main symbol, a structure through which they were evidenced the advances in the design and construction of Paris and, which was established as a manifesto of faith in the commitment to technical progress [4] evidencing the importance of the application of physical concepts in architecture. The Eiffel Tower project was initially planned under the temporary duration of the universal exhibition, however, its construction system, its technological advance, hierarchy, and the successful implementation process next to the Seine River and the Champs de Mars made this project it will last in time and become a landmark, an urban heritage of humanity [5].

In this sense, this research posed as research questions: what is the application of physics in the design and construction of architectural projects? Can the application of physical concepts favor the structural and morphological design of architectural projects? And is it important that architects in training (students) recognize the basic concepts of physics and its application in architecture? The development of this research made it possible to demonstrate the importance of implementing physics within the design and construction processes of Architecture, highlighting that the relationship between physics and architecture acquired greater relevance and meaning from the industrial revolution. For this reason, its study, analysis, understanding, and application are essential elements within the teaching-learning processes of architecture.

2. Method

The methodological framework of this research work was developed from a documentary study [6] that established the importance and relevance of the Eiffel Tower for the universal exposition of Paris in 1989 and the technical aspects derived from the physical concepts applied to its design and construction. Likewise, a descriptive quantitative approach [7] was used that allowed the recognition of perception and knowledge of the basic concepts of physics implemented in architecture by gender differentiation, under the application of the survey as a research instrument and data collection from its primary source [8], its population focused on third-semester students of the Architecture of the Universidad Francisco de Paula Santander, Colombia, and the sample corresponded to 40 students.

For the selection of the informants, a non-probabilistic convenience sampling [9], was used through which the physical and architectural concepts of the Eiffel Tower were investigated among the students who were enrolled as students by the first semester of 2021 active, and that according to the curriculum of the architecture program they were finishing the basic cycle of academic training. The survey instrument was designed ad hoc by the researchers, it was composed of five questions posed under the scheme of the Saber Pro tests; that is, contextualized questions with a multiple-choice option with a single answer. The instrument incorporated supporting images about the Eiffel Tower into the sentences in order to provide visual support for reasoning and the subsequent selection of the correct answer.

3. Results

The Eiffel Tower challenges the physical concept of gravity from a false instability and a height corresponding to 300 meters, which classified it as the tallest element in the world for 41 years, after the inauguration of the Chrysler Building in New York that surpassed it by 19 meters and, later by the Empire State Building that surpassed it by 81 meters in height [10,11]. The structure of the Eiffel Tower is made up of four arched bases that make up a 275-meter-high column and three levels that act as dividing platforms.

The great structural challenge of the tower consisted in the very weight of the structure and the handling of the wind. Regarding the weight, the form of an intensive solid is adopted, this refers to the fact that each section of the solid is subjected to the same tension. As can be seen in Equation (1), it is verified that the pressure at any height of the tower is constant; next, by clearing the denominator of this expression, Equation (2) is obtained. This when solved as a differential equation by the method of separate variables, results in Equation (3), through which the statement is ratified, given that it is a function exponential.

$$k = \frac{V_x}{S_x} = \frac{\pi \int_x^{\alpha} [f(t)]^2 dt}{\pi [f(x)]^2}, \tag{1}$$

$$\int_x^{\alpha} [f(t)]^2 dt = k[f(x)]^2, \tag{2}$$

$$y = a * e^{-bx}. \tag{3}$$

Regarding the effect of the wind, the structure was designed in such a way that, for each elevation, the maximum moment generated by the wind is compensated by the moment of the weight of the tower (see Figure 1). To achieve this balance, the Eiffel Tower has curved legs, in such a way that its tangents, drawn at points at the same height, always intersect at the point through which the resulting wind effort passes, on the part that is above the analyzed points (equality of moments in Equation (4)).

$$F_v * h = W * b. \tag{4}$$

The structural design of the tower was largely elaborated from mathematical calculations applied to the arches, which formed the base, and from the implementation of the physical concept of aerodynamics [12], since the aerodynamic resistance is of vitally important to ensure structural integrity.

Figure 1 shows that the formal aspects that accompany the design of the Eiffel Tower were conceived from the configuration of an open and self-supporting structure [13], which implemented design concepts such as symmetry; understood in the project as a balanced structural, formal and visual arrangement [14], the rhythm; established from a prefabrication process but not standardization, because the variations in the sections of the pieces and the curvatures of the vertices, did not allow the unification or the interchange of pieces [15], a square pyramid shape with a base of 124.90 meters and a height of 324 meters that affected the balance and the distribution of live and dead loads due to their weight and height [16].

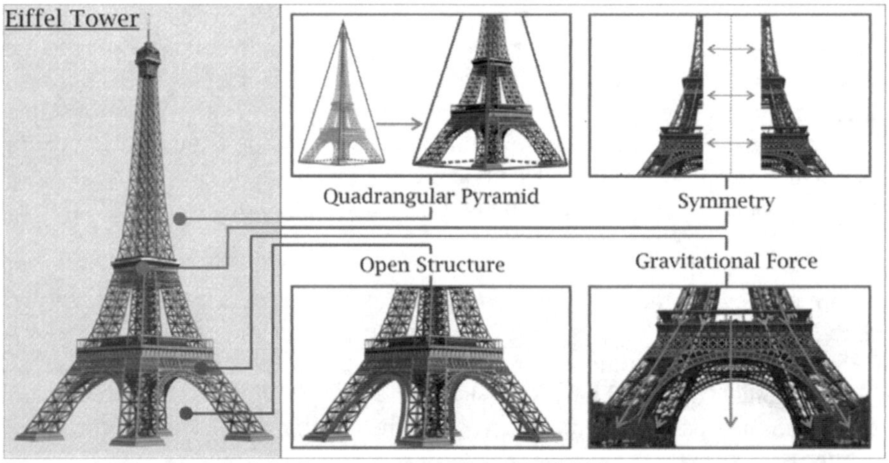

Figure 1. Physical and design concepts of the Eiffel Tower.

From the structural aspect, it should be mentioned that the tower presents wind and gravitational loads [17], as well as compression forces in the arches and compression-traction in the metallic triangular framework, in addition, a distribution of internal forces from the stress of torsion, represented in the internal rotation of the pillars that join the rest of the metallic structure [18,19]. In Figure 2 it can be seen that the tower implements the concept of aerodynamics, which allows the circulation of the wind through the structure [16], finally, it was established that the Eiffel Tower manifests a constant tension throughout its height, which it allows reducing the upper sections with respect to the lower ones and also, that from the beams tension can be generated in the upper part and compression in the lower part of the structure [20].

The Eiffel Tower is presented as a significant reference that allows making visible the importance of the use of physics in architecture, by considering that the implementation of physical concepts can strengthen the design, structure, and aspects derived from construction in architecture; for this reason, the present study considered it important to recognize the level of perception and understanding manifested by students of the third semester of the architecture program with respect to the basic concepts of physics applied to architecture.

In this sense, according to the demographic characteristics of the informants, it was possible to identify that 59.1% corresponded to the female gender with an average age of 19.6 years, and 40.9% to the male gender with an average age of 20.1 years. Regardless of gender, 71% come from stratum two and 29% from stratum three, residing in the San José de Cúcuta, Colombia, and its metropolitan area. This research was organized as a comparative study based on the gender of the informant that allowed to recognize the perception and knowledge of the basic concepts of physics with respect to the Eiffel Tower.

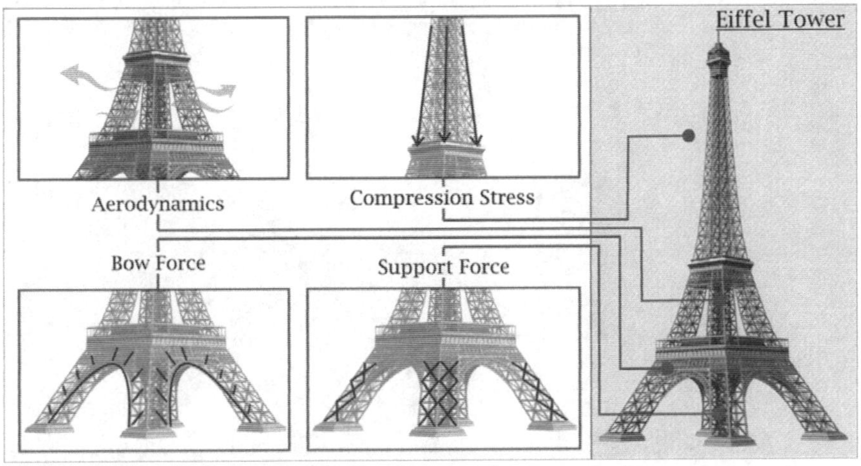

Figure 2. Physical concepts in the structural component.

Regarding the question, what physical concepts are present in the design of the structure of the Eiffel Tower? It was determined that 4.5% of the sample (equivalent to 7.7% of the women) assured that the structural design of the tower presents physical concepts such as: malleability, reciprocal quantities and the Avogadro principle, a situation that evidenced ignorance of the subject, since Avogadro's principle is associated with the ideal gas laws, the reciprocal quantities are recognized as scalars and whose product is the unit and, the malleability that corresponds to a property of metals to assume particular shapes, this could be associated with the structure but it differs, in the sense that beams of different sizes were incorporated into the Eiffel Tower.

Regarding the physical concepts associated with the structure, it was identified that 13.6% of the sample (with relative percentages of 15.4% women and 11.1% men, respectively), assured that the concepts of rhythm, transmissibility and capillarity, were the ones that influenced the structure of the Eiffel Tower, which shows a certain level of knowledge of physics, since the principle of Transmissibility is understood as a force applied to a rigid body, which could be replaced by another force with equal intensity, meaning and direction of the original force; but the concepts of rhythm and capillarity, are not located as physical concepts but design.

81.8% of the students in the sample recognized that the concepts of tension, aerodynamics and torsion were the most influential in the design of the object of study, understanding that a mechanical stress is exerted on the structure, and that its aerodynamic shape, it facilitates the flow of the wind throughout the structure, as well as that the torsion is present in the form that the structural elements take (see Figure 3).

The survey provided to the students the definition of the concepts of hierarchy and asymmetry in order to identify which of them is associated with the design of the Eiffel Tower. It was determined that 86.4% of the sample considered that the hierarchy is evident in the monumentality of its form. This response was provided by 92.3% of the total women, while the percentage of men was lower than this value by 14.5%. The remaining percentage chose to ensure that the concept associated with the design of the tower was asymmetry, showing that it was men who most favored this response option, surpassing women in more than 15% of cases.

Regarding the concept of aerodynamics related to the shape and structure of the Eiffel Tower, it is highlighted that all the students surveyed defined that the design of the tower facilitates the movement of air in its interaction with the structure, allowing the circulation of the wind of the place and, in consideration of its height and weight.

In reference to the types of efforts present in the Eiffel Tower, 68.2% of the informants selected compression and traction, from the gender perspective, it was observed that 84.6% of the women chose this answer, compared to 44.4% of men, which made it possible to show that in this item women were more assertive than men. Finally, the type of effort present in the curved shape of the pillars of the tower was identified, for which 63.9% of the informants opted for torsion, with 76.9% of the women recognizing this option compared to 44.4% of men.

Figure 3. Influential physical concepts at the Eiffel Tower.

4. Conclusions

The research allowed from the documentary study to recognize the historical importance of the Eiffel Tower, as a representative and patrimonial element in the urban scale of Paris, which represented the advances in the design and construction of new structures that bet on the technical progress derived from the industrial revolution, which consequently managed to change the urban landscape of cities. Likewise, the documentary study determined the bases and the identification of the concepts associated with physics that were part of the analysis of the design and construction of the Eiffel Tower.

Regarding the research questions, what is the application of physics in the design and construction of architectural projects? and can the application of physical concepts favor the structural and morphological design of architectural projects? Through this research it was possible to establish that the Eiffel Tower as an object of study allowed to understand the application of physics in architecture from the implementation of the basic concepts derived from these branches of knowledge, through the understanding of the complexity of the structure of the Eiffel Tower and the handling of the winds. The above showed that the application of physics can favor the structural and morphological design of architectural projects, which in the case of the Eiffel Tower was evidenced in the application of ordering elements of design such as symmetry, hierarchy, and rhythm and, basic physics concepts such as aerodynamics, torsion, compression, traction, and tension; correctly balanced within the conception of the architectural design and the structure of the Eiffel Tower.

Regarding the research question, is it important that architects in training (students) recognize the basic concepts of physics and their application in architecture? It was confirmed that it is important for architecture students to recognize the usefulness of the application of physical concepts in architectural works; this when considering that its implementation can favor the formal, morphological, and structural aspect of architectural projects. Likewise, through the results obtained from the perception study applied to the third semester students of the Architecture program of the Universidad Francisco de Paula Santander, Colombia, it was possible to show that women present a greater recognition of the basic concepts of physics implemented in the architecture. The above when considering that the percentage of women who identified the correct answer was higher than that of men in four of the five items evaluated.

According to the results obtained within this research, it is recommended that the academic program review the curriculum with respect to the possibility of including a specific subject for the area of physics, as well as evaluate the topics addressed from the structures subjects (I and II), in order to potentiate the recognition, understanding and application of basic knowledge of physics in architecture in order to strengthen urban-architectural proposals from the technical, technological, material and formal components, as well as the competences of the sub area of technique present in the profile of the graduate architect.

References

[1] Chaves J 2004 Desarrollo tecnológico en la primera revolución industrial *Norba Revista de Historia* **17** 93
[2] Otero M T 2007 Los grandes eventos como acciones de relaciones publicas del Estado: las exposiciones universales *Revista Sphera Pública* **7** 175
[3] López I 2014 La aportación estructural del Crystal Palace de la exposición universal de Londres. Una ampliación del enfoque histórico tradicional *Rita* **2** 76
[4] Hernández P 2019 *La Huella de las Exposiciones Universales en el Trazado Urbano* (Madrid: Universidad Politécnica de Madrid)
[5] Aguiló M 2017 Coyuntura la Torre Eiffel apogeo del hierro, símbolo de Paris *ROP* **3583** 6
[6] Peña T, Pirela J 2007 La complejidad del análisis documental *Información, Cultura y Sociedad* **16** 55
[7] Hernández R, Fernández C, Baptista P 2014 *Metodología de la Investigación* (México: McGraw Hill)
[8] Otzen T, Manterola C 2017 Técnicas de muestreo sobre una población a estudio *Int. J. Morphol.* **35(1)** 227
[9] Argibay J C 2009 Muestra en investigación cuantitativa *Subjetividad y Procesos Cognitivos* **13(1)** 13
[10] Bernabeu A 2019 *Estrategias de Diseño Estructural en la Arquitectura Contemporánea. El trabajo de Cecil Balmond* (Madrid: Universidad Politécnica de Madrid)
[11] Lahuerta J J 2008 Empire state building *Minerva* **7** 20
[12] Gallant J 2002 La forma de la Torre Eiffel *Am. J. Phys.* **70(2)** 1
[13] Capriata P A 2016 *Robert Delaunay y la Torre Eiffel. Fragmentación e Interpenetración del Espacio Pictorico y Arquitectonico* (Barcelona: Universitat Politècnica de Catalunya)
[14] Murcia C 2011 *Física para Arquitectos. Apuntes* (Tunja: Universidad Santo Tomas)
[15] Seitz F 2001 *La Tour Eiffel. Cents Ans de sollicitude* (Paris: Ëditions Belin)
[16] Serrano-López R, Mínguez-Algarra J, Cambronero-Barrientos F, Saldaña-Arce D 2014 La adaptación de la forma a las cargas en el arte estructural: desde las pilas y el arco, a la torre *Informes de la Construcción* **66(534)** e021:1
[17] Bueche F J, Hecht E 2000 *Física en General* (México: Mc Graw Hill)
[18] Allen E, Zalewski W 2009 *Form and Forces: Designing Efficient, Expressive Structures* (United States of America: John Wiley and Sons)
[19] Engel H 2001 *Sistemas de Estructuras* (España: Gustavo Gili)
[20] Bertrand L 2002 *Gustave Eiffel* (España: Ediciones Akal)

AUTHOR INDEX